Thousands of worker bees being shaken from a wood and screen package box into a hive body filled with wood frames and wax foundation.

PACKAGE ESSENTIALS
PRODUCTION INSTALLATION MANAGEMENT
FOREWORD BY DEWEY M. CARON

— A WICWAS PRESS ESSENTIALS BOOK —

©2020 Lawrence John Connor. First Edition

ISBN 978-1-878075-61-1

Edited by Randy Kim. Special thanks to the following people for contributions to this book: Dewey M. Caron, Steve Repasky, Charlotte Hubbard, Andrew Connor, Cathy King, and Rich Wieske. Index prepared by Darcy Connor.

Foreword by Dewey M. Caron, Emeritus Professor, University of Delaware.

All photos were taken by the author or under the supervision of the author unless otherwise noted.

Wicwas Press, LLC • Kalamazoo, Michigan • www.wicwas.com

All procedures and methods outlined in this book may change when applied by different beekeepers operating under varied conditions and when using different genetic stocks.

Printed in the U.S.A.

First Edition, first printing

ISBN 978-1-878075-61-1

Cover photos: Front: Plastic package cages that have just been filled with worker bees and given cans of sugar syrup. Queens in cages will be added later. Outside bees will be removed prior to shipping. Inset: Shaking a wooden cage filled with about 10,000

TABLE OF CONTENTS

—FOREWORD—

The reasons for wanting to own a colony of bees are as varied as their keepers—one person may want to provide pollination for their home vegetable garden while another wants to produce an income by selling honey. Once the proud owner of one colony, beekeepers have various reasons to start another, such as pollinating flowers for song birds and game animals or increasing honey sales at the City Market held every week.

If you are thinking of starting a bee colony—for the first time or once again—the purchase of a package represents an excellent means to success, especially when certain basic beekeeping management tools are employed and carefully followed.

But, like much in beekeeping, there is no one correct way to start. Staring a bee colony is a continuous process, and each year the typical apiary will have one or more new colonies depending on the beekeepers' need. Sometimes, the simple act of buying and installing package won't solve every problem in a particular area.

The accurate diagnosis of what might have caused a bee loss can mean the difference between success and failure. The cause for loss could be one of several issues related to queens or the lack of proper attention once the bees are installed. Or it might be the bees failed to store adequate resources or the colonies were subjected to an exploding infestation of varroa mites from neighboring hives. It is not easy to plan for these losses.

Despite these potential and common problems, packages still carry an excellent probability of getting a new colony started. This book, *Package Essentials,* will help each beekeeper increase their chances of success.

Package Essentials is more than just a discussion of how an individual can start their first colony, replace a colony that was lost, or to use package colonies to increase their total colony numbers. The book includes two methods of installing a bee package, presented through a step-by-step photo essay (Chapter 4). The photos clearly demonstrate the standard procedure used by many beekeepers across the country. They take the mystery out of how to best install bees into empty hive equipment and ensure the colony gets off to a good start.

Additional chapters include behind-the-scenes details on how individual producers across the United States put packages together, from fine-tuning bee stock to rearing queens to be caged within the packages. The methods used to produce packages can vary, as explained in the management techniques used by different producers. This information should help package users understand how and why they should handle and monitor package colonies once they are introduced.

Package Essentials details the basic, time-tested starting method of establishing a new colony with the purchase of package bees. This book will help flatten the learning curve of starting a bee colony. Discussions are clear and concise. Readers will like the discussion of the advantages and disadvantages of using specific bee stocks currently available for the new beekeeper.

Regardless of the choices you make in starting a bee colony, make sure to enjoy the adventure, and best of luck while starting your new colony with package bees.

—Dewey M. Caron, Ph.D.
Emeritus Professor, University of Delaware.

—Introduction—

For years, new beekeepers have started their craft by obtaining a swarm or installing a package of bees. Famous authors and commercial beekeepers started when a swarm of bees landing somewhere in their daily life. For jeweler A. I. Root of Medina Ohio, it was when a swarm landed at his shop one day. Root went on to manufacture beekeeping equipment, form the Root Candle Company and start *Gleanings in Bee Culture*, now called *Bee Culture*.

Other beekeepers have their swarm stories. For Professor Alphonse Avitabile, it was a swarm that landed on the sign for the University of Connecticut branch campus in Waterbury, Connecticut. He went on to study bee behavior and published research on wintering colonies in southern New England. He also collaborated with Dr. Diana Sammataro in writing *The Beekeeper's Handbook*, a pillar of beekeeping training for thousands of new beekeepers.

Package bees and nucleus colonies are an alternative to life-changing moments when a swarm unexpectedly enters our lives. If we start with the purchase of a nucleus hive or a package of bees, our first beekeeping work will be easier to manage. It has been extremely rewarding that one of my most popular books is *Increase Essentials*, now in its second edition. While I prefer to start with a nucleus colony, they are not always available. In fact, most new beekeepers have just one option: package bees.

Starting with a package colony—or more as I recommend—is a relatively easy thing to do. While not inexpensive, packages are a cost-effective method of obtaining colonies. So, if you are new to beekeeping, this book provides an overview of how package bees and queens are produced, how to install them, and how to manage them until they are fully established in your apiary.

—*Lawrence John Connor, Ph.D.*

—Chapter One—

Packages and Swarms

Both swarms and package bees are used to start new colonies, and they share a large number of traits.

Swarms are natural—nature's way of making more bee colonies—but the timing of their emergence is often a complete surprise. While their biology is similar to package bees, the exact moment of swarm production is rarely predicted accurately. Packages, often considered artificial swarms by many beekeepers, have a highly predictable delivery date. The use of packages is an excellent way to establish colonies. While swarms force you to "Bee Prepared" and have swarm-catching equipment with you at all times, packages simplify the process considerably.

Strong colony of Russian bees issuing a swarm. It was early June, a common time for colonies to reproduce. About 12,000 worker bees and one or more queens leave with such a swarm and fly to a new nest site.

Bees are not the only insects that swarm. Other insects, such as termites and ants, swarm when the parent colony in the original nest generates reproductive individuals to establish new colonies. With both termites and ants only the reproductive males and females swarm. The non-reproductive workers are left behind. Such swarming is more of a sexual reproductive event, not an asexual colony-level reproduction activity as found in honey bees.

Within the Genus *Apis*—which includes our Western Honey Bee—worker bees, not reproductives, make up the majority of the swarming individuals. Usually just one mated queen or up to a dozen un-mated reproductive queens, called virgins, leave with the swarm. Any male bees, or drones, seem to be along for the ride and are not specifically present for the purpose of mating with un-mated queens in the swarm.

One major difference between swarms and packages is that packages are human-made. They are supplied by package bee producers, specialist beekeepers who put a great deal of time and money into producing lots and lots of bees.

Packages are sold by the weight (pound or kilogram) of worker bees. These worker bees are usually shaken from the brood nest of strong hives. Because these bees are young they will live for several weeks and will work hard to grow and develop a new colony once the package is installed. This selection and use of young nurse bees in package colonies is a key part in starting a new colony.

For package producers, honey production is usually secondary to bee production, although most package bee producers produce surplus honey to build up the colonies for the next season. Season wide, naturally collected pollen and nectar are superior for colony growth compared to human feeding with sugar syrup and protein supplements. That is why beekeepers incur considerable expense to move their package-producing colonies to high-quality honey producing regions of the country.

Earlier I wrote that the timing of package production is more predictable than swarm production, but unfortunately nature often interferes with package production colonies and their ability to build bee populations. Cold weather, heavy rains and inadequate bee forage are key concerns for package bee producers.

Swarms and packages share several key aspects:

(1) Both consist of mostly young worker bees that are needed to establish and grow a new colony in a very short time period—as little as 90 days during ideal forage conditions.

(2) Both have a queen, mated our un-mated. Packages queens are mated as standard practice.

Large swarm in Hawai'i in a bivouac or temporary location while the scout bees decide where to move the bees for a new, permanent location. Swarms are unable to move beeswax honey comb. L. Rusert.

(3) Both travel with food. Bees in swarms carry honey in their honey stomachs while packages are shipped with a container of sugar water.

Both swarms and packages are alike in what they don't have:

(4) Neither leave their parental hive with nectar, water, bee bread (fermented pollen), brood (developing bees—eggs, larvae and pupae), or honeycomb (hexagon-shaped beeswax storage facilities). Both swarms and package bees must produce beeswax comb after the bees become established their new home.

(5) It takes Western European honey bees 21 days to complete their metamorphosis from egg to new adult. The average life expectancy of young worker bees is four to six weeks, but some die sooner than that. If the swarm or package contain older bees, they will start to die within a few days and may all be dead by three weeks, before any

Swarm of bees being shaken into an unoccupied hive body with empty comb.

new brood emerges. In extreme cases, this may cause a population collapse. This is why good package producers never shake bees from honey frames, where older workers are likely to be located.

This age gap is a critical aspect of development in both packages and swarms. The 21 days it takes for packages or swarms to produce the very first emerging bee means that all the original worker bees in the colony will be at least 21 days old or older when the first new bee emerges.

As a part of the complex division of labor in the hive, bees perform many different duties as they age. This means that by the end of an older bee's life, she will have performed many diverse chores while helping establish a new colony. They have raised brood, built beeswax, processed nectar into honey and pollen into bee bread, guarded the hive, run off intruders, and foraged for pollen, nectar, propolis and water. These old bees often die with tattered wings and bodies worn naked of their body hairs. Worker bees just wear out during the spring, summer and fall. In the winter, when they are not foraging and there are fewer brood cells to feed, worker bees often live for the long winter months.

In both swarms and packages, the death of large numbers of individual bees means that there is a decline in bee population at

Wood and screen cages holding California packages of worker bees and one queen per cage. These bees were transported to Kansas City for distribution.

Beekeepers shaking bees out of the shipping cage and into the space of an empty hive body, much like shaking a swarm into an empty box.

the exact time when the amount of brood should have expanded considerably. Fortunately, this is usually not a problem, but if it is cold and rainy, or there are fewer flowers to visit, or if colonies are exposed to stressors like bee diseases, pests and pesticides, bees fail to thrive and some will die. These stress-related issue make it difficult for a new colony to grow and prosper and for the beekeeper, very challenging to manage.

Through natural evolution, swarms expand colony numbers and are successful in increasing the number of thriving bee colonies. Swarming is not a form of sexual reproduction but a form of colony-level reproduction and how colonies make new daughter colonies. Both package bees and swarms share risks. They both may experience a wide range of problems—concerns we will explore further in this book. Our goal is to help the bees so that they do not end up in failure. Humans can play an important part in increasing colony success, especially with swarms and package colonies.

Keep in mind that a colony's production of a daughter swarm gambles with colony resources in an effort to produce a daughter hive. Both the parent hive and the swarm they produce have risks because swarming is far from being a predictable process. Survivability is based on genetic fitness, the availability of food resources, freedom from diseases, parasites and pests, and old-fashioned luck. Research has shown that the chance of a swarm reaching its first birthday is less than 20%.

Package bees are a human attempt to improve natural selection's success rate. Nature's success benchmark is relatively low, so if you get more than 20% of the packages to survive, you have improved nature's survival rate with swarms. Most new beekeepers expect a 100% success rate with package bees, but by the end of the first year, the actual number is usually much less. Because package bees cost a lot of money, any loss is painful. Throughout this book we will look at ways to decrease losses and increase colony success.

THE BIOLOGY OF PACKAGES AND SWARMS

If you are new to beekeeping and starting new colonies, you will benefit with a little knowledge about package and swarm biology. We need to discuss the division of labor in a bee hive. When a package or swarm is installed or reaches a new home, most of the bees are relatively young, and the colony divides the work effort. The older bees often become the new field force of scout bees and foragers. They gather pollen, nectar, propolis, and water and carry it to the new nest, while their younger sisters process and preserve this food and later digest it to feed their immature sisters and brothers. Younger bees also build the much-needed wax honey comb, serve as care givers to the queen, and care for the newly produced larvae.

Table One: HOW SWARMS AND PACKAGES COMPARE

Characteristic	Swarm	Package
Number of bees	8,000 to 16,000	10,000 for three pounds
Food reserve	Honey in stomach	Syrup in can
Queen	Mother or sister	Unrelated
Timing	Bees determine	Shipper determines
Disease risk	Equal to local bees	Usually medicated
Mite risk	Equal to local bees	Often treated
Brood and comb	None	None
Success rate	Less than 20%	More than 20%

When swarms leave their parent's nest, they leave without furniture—the wax honeycomb—which serves as both the bedchamber and the food pantry. The same is true with bees used to form a package of bees. But the bees that leave in a swarm carry inside their abdomen a honey stomach full of honey, gathered from the parent hive's food pantry. They use this honey to produce beeswax and brood food to feed the queen and keep the population of bees alive. Bees shaken into packages, on the other hand, are not given a chance to stock up with honey from the hive's pantry before they are shaken and shipped, so they travel with a can of sugar syrup.

Almost without exception the worker bees and the queen or queens in a swarm are genetically related. All the bees in a hive that have the same mother queen are called sisters. If they also share the same father, we call them super-sisters. You cannot usually tell them apart by behavior, but sometimes you can see different color patterns.

This is one place where packages and swarms differ dramatically. In a package, rarely is the queen the mother of the workers. She may not even be genetically related! The reason for this is pretty simple. Beekeepers who produce package bees are under a lot of pressure to produce and mate large numbers of desired queens for beekeepers across the country. These queens are mated and placed with two- or three pounds of bees obtained from whatever colonies were next up on the producer's shaking schedule. The bees in a package may have come from many different colonies. They are not sisters and are often completely unrelated.

Consider what happens with a commercial beekeeper's shaking crew. They harvest worker bees from colonies located in remote locations, set up so the bees are allowed to forage on spring flowers and grow in bee numbers, providing the maximum number of young bees. The shaking crew may visit these colonies and shake bees just two or three times each spring, harvesting young worker bees from the broodnest, but leaving the queen, the drones and the older foragers back in the hive to continue growing the colony.

Does it matter that the queen and the workers are not related? The answer is maybe. It might matter when the queen is quite different genetically, coming from a different subspecies or race (there are

Large apiaries are used to produce bees for packages. Many of them were established from package colonies themselves.

29 different races of *Apis mellifera*, a huge diversity). Fortunately, in most cases, it is not a critical factor. In fact, many beekeepers enjoy seeing the worker bees change color when the daughter workers produced by the new queen start to emerge and replace the founding workers. This is particularly dramatic when yellow/gold worker bees travel with a queen that produces black or grey/black worker bees or the other way around. There may be shifts in behavior as well, such as the frequency of stinging behavior. Color and behavior shifts may go either way and are usually often noticed by the new beekeeper.

Luck determines much about the fate of bees, including natural swarms or human-made package colonies. Most beekeepers end up selecting a combination of swarms and packages as a means to replace losses and increase colony numbers. For new beekeepers as well as those expanding their operation, the choice is usually between buying package bees or obtaining nucleus colonies that contain a laying queen with her own brood and bees. Nucleus colonies are discussed in detail in *Increase Essentials*, a companion book to this one.

The absence of brood and the potential of a population gap are not usually issues with colonies that have been made by splitting—such as increase colonies made with the Doolittle Method. These increase colonies (splits) are the main difference beekeepers have to consider when making new colonies when compared to package bees.

A frame containing dead bees and brood This equipment may be used to establish a new package colony. The bees will clean out dead bees and any moldy frames.

WHY USE PACKAGES?

Beekeepers use package colonies to establish new apiaries or replace lost colonies. Some beekeepers use packages as a lower-cost method of increasing colony numbers. Traditionally, package colonies have permitted beekeepers an economical means of setting up and installing new bees on new equipment.

Packages are easy to obtain in the spring (if you plan ahead), accessible and consistent (depending on the seller). Your local package bee distributors and re-sellers are established and respected parts of the local beekeeping community and form a 'package deal' of support, mentoring, education, supplies and bees. New beekeepers should seek this 'package deal', and buy from a local supplier who is connected to the entire whole bee community.

In the 1950s, my family ordered package colonies from the Sears Farm catalogue to establish colonies. This year, I obtained one package from California to replace a colony I lost during the previous season. The packages were produced by the Olivarez operation in California (OHB Inc.) and distributed by the local Dadant and Sons branch manager located less than one hour away from me.

It installed my package bees on the comb and honey from the colony that died. I placed the queen in her cage between two drawn frames. The worker bees were attracted to her pheromone or scent and crawled out to surround here, thereby emptying the cage. The bees cleaned out cells and reworked the old combs and did a great job. In a matter of a few days there was no evidence of the old dead colony.

• Using Packages to Replace Lost Colonies

Both natural and managed bee colonies fail, and there are several reasons why this happens. Some beekeepers think colonies die during the winter because of cold-related starvation, while others think they die from queen failure, poor food supplies, pesticide exposure, varroa mites, small hive beetles, bee diseases, or several other reasons. Of course, sometimes two or more of these factors contribute to colony death.

This colony died during the winter. The beekeeper has plugged all entrances to prevent the potential of disease spread if American foulbrood is present. When tests showed no disease was present, the equipment was used to establish a new package colony.

Installing package bees is a simple, effective and time-tested method to replenish dead colonies. By knowing where my package dealer obtained the bees and that the queen was from mite-tolerant stock, I reduced my colony's varroa mite problem potential. I was able to reuse combs and use stored honey from the dead colony, which is a huge boost for a new package of bees. The colony built up so fast that I split it three months after installation.

• Using Packages Where Nucleus Colonies Are Not Available

In certain parts of the world, especially in cold climates, package bees are a clear option to keep bees. In Alaska some beekeepers overwinter their colonies while others intentionally kill their bees in the fall, clean up the equipment and restock the hives with packages the next April. They must balance the high retail value of Alaskan fireweed honey with the loss of leaving a great deal of honey on the hives, heavily feeding their bees, or restocking the hives in the spring. So some Alaskan beekeepers choose to destroy colonies in the fall as a cost-effective way to harvest and sell more honey. While their production season is short, the long summer daylight keeps bees foraging 20 or more hours a day, providing a greater opportunity for large honey yield.

While U.S.-produced packages were traditionally utilized by Canadian beekeepers, the movement of bees, comb and brood between the United States and Canada has been banned since 1987. With the discovery of mites in the United States, Canada closed its borders to U.S.-produced packages to keep parasitic mites out of the country. Even though the mite species eventually made it into Canada anyway, the U.S. borders remain closed to both packages and colonies. The border was opened to U.S.-produced queen bees in 2003.

Some Canadian beekeepers have invested in elaborate inside wintering facilities while others import packages from the southern hemisphere producers. Only a few produce package bees in the few limited micro-climates of the country where temperatures support earlier colony development.

• Using Moldy Frames Filled With Dead Bees

Old, wet, and moldy combs filled with dead bees can be shaken onto the ground or brushed to remove surface bees. I was able to add frames of honey to accompany these combs. Once installed, the bees from the package rapidly dried out the combs and removed dead bees and debris from comb cells rather quickly. They are amazingly efficient. In a matter of a few days after the package was installed, the combs were renewed, and the queen had laid eggs that developed into larvae in many of the cells. Other cells contained freshly collected nectar and pollen from flowers, being converted into honey and bee bread.

• Determine If Disease or Pesticides Were Involved

Dead colonies require extreme caution, especially if American foulbrood or pesticides are involved. Never use combs from a colony that died from American foulbrood. This is a stubborn, deadly, spore-forming bacteria. Potentially contaminated equipment must be quarantined immediately and either burned or irradiated. Otherwise, the spores survive and not only kill your colony, but potentially kill every colony housed in that hive afterward as well. It poses a huge risk to all the colonies in the area. AFB is serious business!

Detecting pesticide contamination must be based on the dead colony's history. If there is the potential of loss from chemical contamination, the combs containing brood and bees should be removed and rendered for wax, or better yet, burned or buried in the ground. Honey frames may be used to establish new colonies, but only a frame or two at a time to determine any lingering chemical residue.

Package bees provide a predictable, lower cost method of starting, replacing or growing colonies. Once a beekeeper learns to provide a balanced bee population, provide proper food reserves, and watch for queen failure, diseases and pesticides (which all seem like a lot to learn), packages are a good system. It takes some training and a little experience to be successful with package colonies, but they work and can reward the diligent beekeeper with a productive and vibrant colony for years and years to come.

—Chapter Two—

Package Bee Production

Package bees ready for shipment to customers. They are spaced on pallets and positioned to allow for maximum airflow inside a ventilated, dark holding room.

Package bees are human-made shipping boxes filled with a mated queen, several thousand worker bees, and sugar syrup to feed and hydrate the bees during travel from the package producer to the beekeeper.

The bees in a package resembles a resting bee swarm. Within each box workers move freely, but the queen is confined inside a special cage and provided with a bit of sugar candy for food. Workers feed her through the screen of the cage and obtain her pheromones. Once introduced into a hive, this chemical communication stimulates the bees to develop a strong and healthy colony.

To produce each of the components of a package, bee producers maintain two independent but highly coordinated beekeeping operations: queen production and bulk bee production.

Queen production and rearing are managed by a queen crew, while the worker bees are mass-produced in production hives carefully managed by a team of beekeepers called the shaking crew. Both operations are crucial to the production of packages, so let's examine both in detail.

QUEEN PRODUCTION

Queen production may last throughout the beekeeping season or for just the spring months, depending on climate and beekeeper demand. Regional differences affect the start of the queen-rearing season. In warmer areas like Florida, the queen production season begins as early as January, while in northern California production begins around March 1st following the almond bloom. Climate change impacts production cycles through changing temperature, rainfall and forage availability.

In tropical Hawai'i where the weather is good and food is available year-round, the production cycle is determined by customer demand. Several large package bee operators produce queens around Kona to supply queens for mainland packages produced thousands of miles away. These early queens come with the added

Beekeeper on Hawai'i's Big Island feeding colonies involved in queen cell production. Several large producers ship tens of thousands of queens per year.

risks associated with shipping delicate cargo over a considerable distance. Such risks including overheating, chilling and the exposure to insecticides. Queens produced on the tropical island's resource-rich environment avoid many issues that occur on the mainland, such as inconsistent and unpredictable spring weather. Tropical queens have been shown to be well-mated. They may lack genetic adaptations for survival in cold climates, however.

Queen producers located on the continental U.S. often find that weather is the key issue that affects both general queen production conditions as well as mating flights. Warm weather stimulates the production of nectar and early pollen species (willow, maple, alder) while cold weather suppresses food production, causing a delay in mating that will inevitably push back the package bee delivery schedule. Heavy spring rain interferes with colony movement, production conditions and bee forage.

A common challenge for queen producers located in areas of cool, wet or generally unfavorable weather is that queens fail to mate because extreme weather conditions prevent drones from flying outside the hive, resulting in fewer viable mating partners for queens. Drones and young queen bees only mate in the afternoon

The use of queen excluders between hive bodies allows beekeepers to stimulate cell finishing of thousands of grafted larvae.

when the temperature approaches 65-70°F. Anything that interferes with this, such as rain, high winds or cold temperatures, potentially decreases mating success. Fortunately, bees are pretty good at finding any narrow window of opportunity for sexual reproduction. An hour or two of good weather during a rainy period, timed in the afternoon and during low wind conditions, provides enough time for an opportunity for mating to occur.

Queen mating and the eventual package production can be unpredictable. With climate change, we see wide swings in weather conditions. Critical buildup and mating periods may be extremely warm and windy, or they could be cold and wet. Many important pollen and nectar-producing plants may flower too early or too late, interfering with the proper nutrition of package bee production colonies.

Queen cell and mated queen production methods may involve considerable variation from one beekeeper to another. Here are the steps outlined in *Queen Rearing Essentials*:

1. Queen cell producing colonies must be very strong since queens are important to the livelihood of the hive, especially a new one. Beekeepers often supplement these colonies with frames of emerging brood to ensure high bee numbers and increase overall success. I prefer this method over adding adult bees in bulk, which comes with the added risk of fighting between the bees.

There are many commercial queen rearing operations producing package bees. In some areas, all queen production is done locally, such as in this location on a Caribbean island. Here, a beekeeper is grafting larvae to produce queens.

In commercial queen rearing production, grafted cells are positioned on bars. Here three grafting bars are held on one grafting frame. These cells appear to have a 100% 'take' or acceptance.

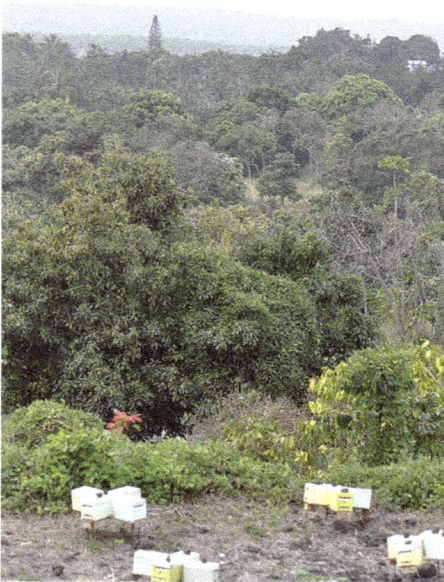

Queen mating yard in Kona, Hawai'i, on the edge of the Pacific Ocean.

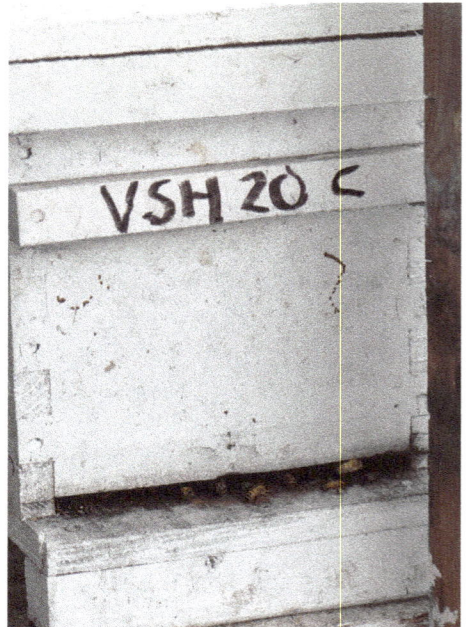

Varroa Sensitive Hygienic (VSH) breeder queen being used to produce daughters to spread varroa mite tolerance.

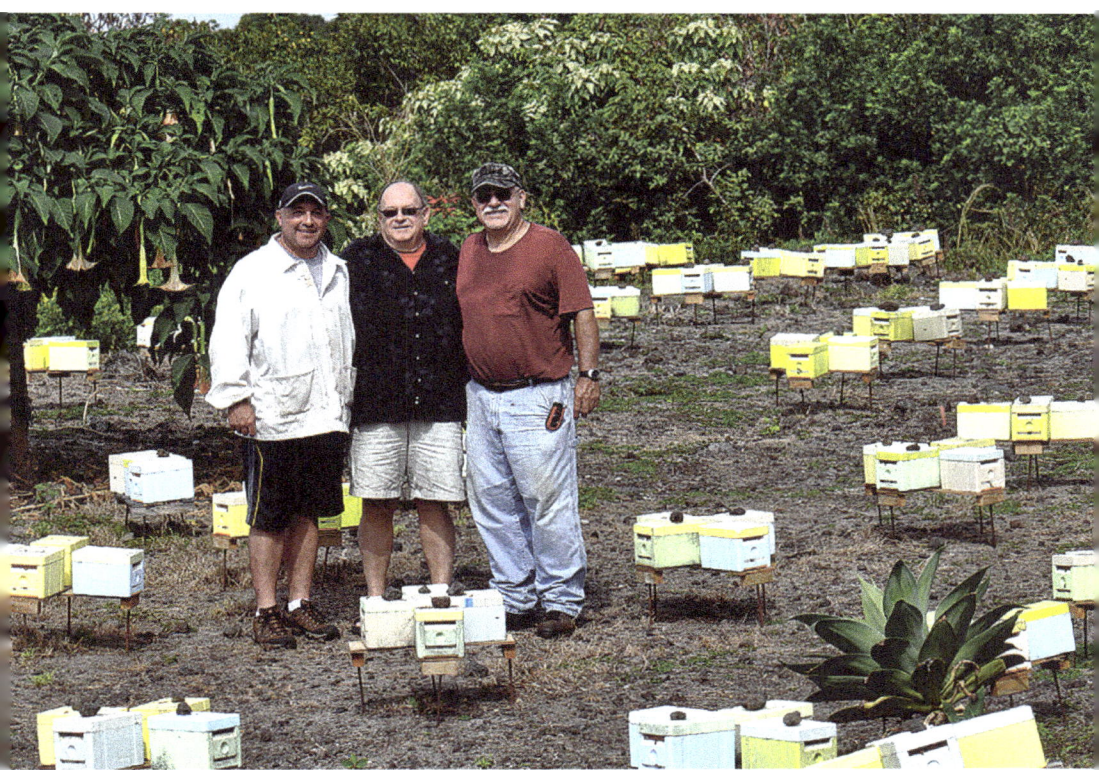

The author (center) with Russell Olivarez (left) and Ray Olivarez Sr. in a queen mating yard in Kona, Hawai'i. Each group of four mini nucs has a metal hive stand that gets them off the lava rock and away from bee-eating toads.

2. Several methods of starting cells are used by queen producers. One popular method includes large 'starter boxes' filled with thousands of young nurse bees. These hives may be closed or allowed open flight, but they are usually set up without frames of brood or a queen.

Starter colonies create a queenless hive environment where there is no queen pheromone and young worker bees are actively producing royal jelly. When young larvae are grafted into cells and are placed into the starter hives, the queen larvae are well fed and the cells are well-provided with royal jelly. Beekeepers usually use sugar syrup and protein to feed the bees, especially in the early season before natural food is abundant or the weather is cooperative. This food is in addition to what the bees gather on their own natural foraging.

3. Once cells are started, they are transferred to colonies called cell finishers or cell builders. These are extremely strong colonies with queens kept under a queen excluder. Nurse bees feed grafted queen larvae located above the excluder next to frames of brood.

In the cell finisher, the bees finish the cell production process that began in the starter hive. As part of their natural biology, bees will build up the cells until they are ripe, containing queens that are within one day of emergence as adult bee.

In the cell builders I set up, I keep a very productive laying queen in a hive body below a queen excluder. I then raise open brood above the excluder, but I keep the queen out. As the nurse bees care for the brood, they are attending to the food requirements of the queen cells, finishing off the cells with both food and a wax capping. Throughout this process, the nurse bees are also playing an active role to ensure optimal physiological development of the queen by keeping the cells at a proper temperature.

Many commercial queen producers use a specialized divider board called a Cloake Board to combine the tasks of starting and finishing colonies into one unit. The basic need for strong colonies with large populations of nurse bees remains.

4. Just before queens are to emerge, their cells are placed into a mating colony, one cell per colony. Most beekeepers use mating nucs, which are similar to an increase nucleus but may be smaller. Many queen producers use one of two standard mating boxes:

Mini 'spam' nucleus with three frames and a feeder. Base of a queen cell shows.

Mini nuc feeder with a few dead bees and small hive beetles.

Mini nucleus in the field in California. The improved thermal properties of the plastic hives stabilizes temperature in these small mating units.

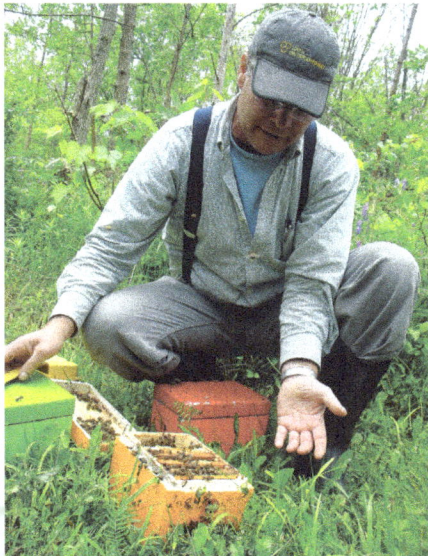

Ontario's Paul Kelley with mini nucs in the Buckfast Bee program on Torah Island.

Frame with bees and mated and laying queen in a mini nucleus. Her body has increased in length and her abdomen is swollen in circumference.

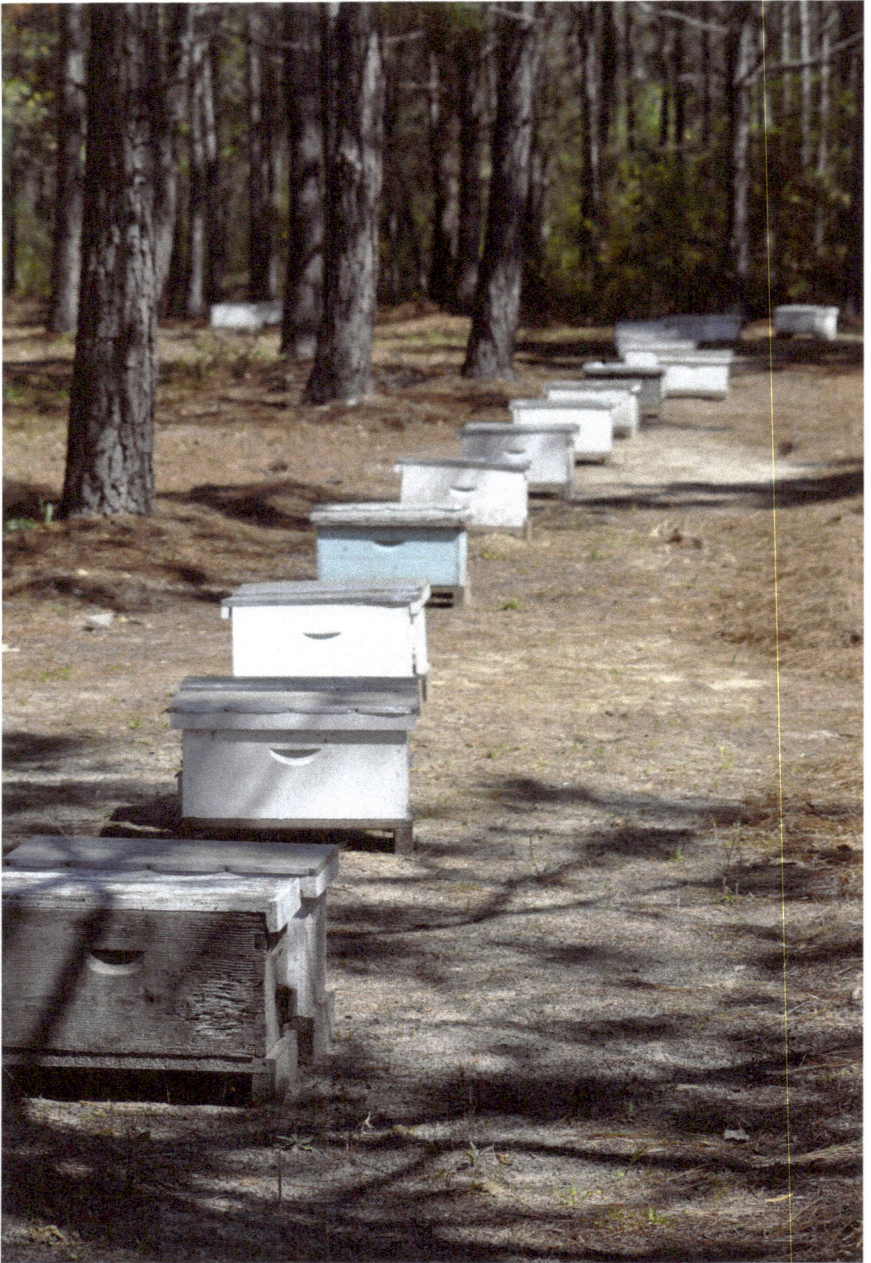

Commercial queen rearing operation Rossman Apiaries in Georgia. There, the small hive beetle has made it necessary to use larger three to five-frame mating nuclei with the advantage of keeping the nuclei in operation year-round.

Compare the much larger bee populations in a full-deep-frame nucleus in Georgia with the mini nucs shown on the previous pages. In the front is a syrup feeder made of wood and coated to prevent leaking.

a. One mating box design is larger and uses standard deep or medium depth frames. This makes the colonies easier to make up as the standard sized frames allows for interchangeability.

In Georgia and Florida, the primary setup is to use 2 to 5 frames in a 3- to 5-frame nucleus box. Some beekeepers regularly use 5-frame nucleus boxes, but they start with only 2 or 3 deep frames early in the season. As the colony grows in size, more frames are added to the nucleus hive.

Frames of brood and workers are located and removed from support colonies. They may be held for several days in screened hives to allow the bees to emerge, mix and settle. These queenless frames of bees and brood are then moved to the mating yard and placed into mating boxes. One ripe queen cell is placed between brood frames, and the colonies are allowed to establish food-gathering and support the queen's emergence, maturation and mating.

These colonies need to be strong enough to combat small hive beetles, a pest first discovered in Florida in 1998.

b. In California where the small hive beetle has not become as well established, perhaps due to the lower humidity, many beekeepers use mini-nucs. These boxes are traditionally made of

Once queens have mated and have been laying eggs for several days they are captured and held in queen-holding frames in queenless colonies.

wood, but a polystyrene version, called a poly-nuc, has proven to be quite popular because it improves temperature regulation and has greater success rate in producing mated queens. The downside is that the frames of poly nucs unique dimensions and not interchanged with other beekeeping equipment. Mini-nucs produce more queens at a lower cost than the larger, full-frame nucs.

These mini-nucs are designed for the mass production of queens in a very short time period. Tens of thousands of these nucs can be produced by shaking bees from the colonies coming out of the almond orchards and subdivided into units of one thousand bees each. Some queen crews refer to mini-nucs as Spam-nucs because they fill an empty Spam can to obtain one thousand worker bees.

When properly fed and supplied with a ripe queen cell, these small units prove to be very effective in producing one or two cycles of queens. The disadvantage is that they do not last more than two or three queen cycles of queens.

The mini-nucs contain two or three tiny combs while the 5-frame nucs contain standard frames.

5. One of the key concerns of all queen producers is the grafting source for the larvae selected to produce queen cells. Some queen

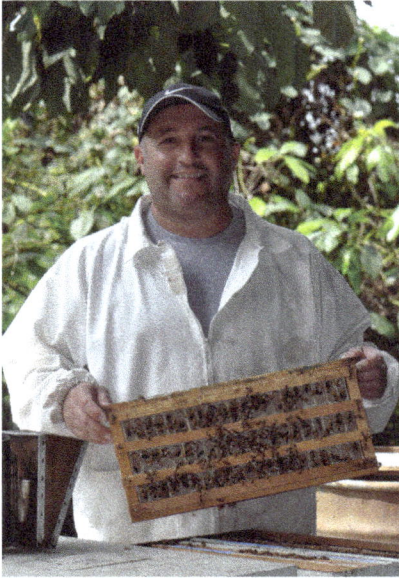

R. Olivarez holding mated queens in a holding frame, about 63 queens per frame.

Each queen-holding frame faces a frame of brood in a queenless queen bank. These banks store queens until shipped.

producers use their best and most promising colonies as grafting stock while others rely on genetically selected, tested and proven queens produced through instrumental insemination. The daughter queens are open mated by the queen producer and harvested from the mating nuc after they start laying worker eggs. These queens are harvested from the nucleus and added to the package and prepared for shipment to the customer. Read more about queen stock in *Best in Stock* in Chapter Three.

MATING

As just discussed, the success of nucleus hives depends on the size and style of the mating box. Large mating colonies are often made up in remote apiaries and carried to the mating yard. The small mini-nucs are made by shaking bulk bees into large wood and screen cages and transporting them to an indoor facility where the bees are measured out into the boxes, fed and provided with a queen.

In a visit to Koehnen and Sons, Inc. of Ord Bend in Northern California, I saw the facility where empty mini-nuc boxes are established in a simple assembly line. Box by box, beekeepers fill the mini-nucs with a full feeder of sugar syrup, insert empty frames, attach a ripe queen cell with a heating iron, and use a Spam-can to scoop bees from the bulk bee container and add them to the nuc. The final beekeeper rubber-banded the boxes and lids together and placed them onto a pallets in a holding room. There they would stay for 36 hours before being moved to the field.

The holding room is dark and well-ventilated. This kept the bees quit and allowed the queen to emerge from her cell and for the worker bees to sense her pheromone, stimulating her proper care and feeding. It also gave the worker bees time to feed on the sugar syrup and build new beeswax comb. Any surplus syrup was stored in the new comb.

In a day and a half most of this will be finished. Then the mating nuclei were moved to the mating apiary and arranged so each hive faces a different orientation than the ones next to it. This helps minimize queen drifting.

While the box is closed the bees received air through a screened ventilation hole in the side of the box. At twilight, a beekeeper goes through the apiary and opens the mating boxes. This allows the bees access to the outside so the bees had the evening to settle, guard their entrance, and orient to their new location during their first flight in the morning.

All these precautions are done to prevent the bees from taking off and not staying in the mini-nucs. There is always some risk that the bees in these mini-boxes will all leave their tiny little homes and form large swarms of bees on a nearby tree, wasting both the queens and the workers!

• Mating Behavior and Hive Arrangement

The location of mating yards should be within flight range of package bee production apiaries to provide an abundant and healthy supply of desirable drones. If packages are produced in a way to remove drones and keep them in the parent colony, the production hives should have a large population of mature drones that fly to special areas called drone congregation areas (DCAs) a short distance from the hives.

Young queens fly a mile to meet with drones from other apiaries, thus preventing inbreeding and increasing genetic diversity.

Queen bank used to move mated queens from the mating yard to the holding and shipping area. A migratory screen allows for ventilation but restricts flight.

Bees have evolved this complex incest prevention mechanism to minimize the number of related drones and queens that mate. Package bees and queen producers supply strong colonies in the general area of queen mating to optimize genetic diversity.

As the price of bees increases and the income from pollination rentals has jumped, beekeepers must protect their hives so they do not lose them to theft. This would eliminate their income from pollination fees, queen production, package bee production and honey production. In package and queen production locations, production hives and mating areas must be protected. Many are located on leased ranch land, forested areas (especially where trees have been harvested and replanted) or protected locations where the colonies are often behind a series of locked gates.

• Letting the Queen Reach Maturity

Holding the queen for maturation is a concern for all beekeepers— both the queen and package bee producers as well as the beekeeper user.

For decades beekeepers pulled queens as soon as they found eggs and larvae in brood cells in the mating unit. Rarely did they wait to see if it was worker brood, indicating the queen was properly mated. Lately there have been scientific reports indicating that young queens do not reach 'maturity' until they are four weeks old, motivating queen producers to leave queens in the mating boxes for a longer time period even though this increases their production time and expense.

Traditionally, queens were harvested on a 12- to 14-day cycle, that is, the number of days from the installation of the ripe queen cell to the time a queen is removed, caged, and shipped to the customer. Now, the trend is to wait until queens are on a 15 to 22 days of age. The extended time gives the queen more time to lay eggs as well as mature physiologically. This may impact the young queen's pheromone production as well as her behavior on the comb, perhaps in ways humans cannot detect. By increasing the time between queen cell installation and queen harvest, queen producers report a reduced number of complaints from customers about queen quality and acceptance.

Extending this time period means a huge shift for the queen producer's schedule. Assuming the number of queens needed by customers remains the same, beekeepers must establish more

After mated queens are removed from mating colonies they are replaced with ripe queen cells—those that will emerge within a few hours. Here, ripe cells are held in an incubator before going to the mating yards.

queen mating units, along with all the support equipment and bee resources and storage space required. For the customer, this means a better queen, but ultimately at a higher price.

• Queen Harvesting

Harvesting queens is not especially difficult in the mini-nucs since there are so few square inches to check for the laying queen's swollen body. When eggs and larvae are seen, and if the queen is large and her body is lengthened and swollen dorso-ventrally (back to belly), then she will be ready to be removed from the mating nucleus.

Larger full-frame nuclei contain more bees and have more square inches of comb. This often increases the "search time" looking for the laying queen.

A huge challenge in any queen rearing operation is the development of queen cells by the bees in the mating nucleus. These so-called "natural queens" or "side-comb virgins" are a headache as they often lack the production condition of larger colonies and desired genetics of the queen cells added by the beekeeper. Their production is off calendar—they would have started from

When not used in package bee colonies, surplus queen bees are sold to beekeepers in bulk packages. Adult bees regulate temperature and support the queens.

young worker larva rather than a ripe queen cell. The beekeeper examining the nucleus must search for the presence of a queen emerged from these cells. The search for queen cells on the surface of the comb with a beeswax flap at the bottom of the cell, as proof that the bees have produced their own queen and ignored the beekeeper's business model.

Restocking nucs with ripe queen cells usually takes place later in the day the mature queens are harvested. Nuclei that need queen cells may be indicated by a stick or a stone on the lid, serving as a signal to the beekeeper to install one ripe cell.

Other beekeepers return the day after harvest and use the feed-can opening in the lid to insert the ripe cells, reducing the disturbance to the bees. A longer wait of the cell's introduction increases the pheromone-free period and increases the production of natural queen cells, making the mating unit potentially unproductive.

ABUNDANT YOUNG BEE PRODUCTION

Colonies that produce bees for packages traditionally are filled with abundant bees, meaning that they would generate large swarms if bees were not harvested from them. These colonies are managed to produce a large number of young workers ready to be removed by the beekeeper and shipped to customers. Old bees do not work well in package colonies—you do not often recruit a vigorous fighting army from a senior center.

Here are some common practices used to produce abundant young worker bees for package colonies:

1. Production colonies that are managed for young bee production possess queens selected for their high egg-laying rate (the number of eggs she produces every day). More eggs mean more bees, which means more available bees to be put into packages.

2. The queens in these colonies are usually less than two years old. This ensures that each individual colony will produce the maximum number of bees. It also reduces the number of colonies that fail.

California-based commercial queen and package bee producer Ray Olivarez replaces the queen at the end of each production colony after the second shaking season. His records show that if he keeps queens for a third year, the operation's per-colony production declines. He also said that, as a result of his requeening plan, he is

Ray Olivarez working one of the colonies used to produce bees for packages. Weak and lost colonies must be replaced and fed as in any beekeeping operation.

the largest user of the queens he produces every season. If there is a problem anywhere with his production system, he would see it in his own bee colonies!

3. These production colonies are also used in honey production and pollination. Referencing Olivarez again: All colonies are moved to almond pollination in February and managed for maximum bee strength to provide the greatest number of foragers on the flowers. Once the almond trees reach petal-fall, the colonies are moved to apiary locations where there is an abundance of pollen and nectar plants. The colonies are often very strong because the pollen and nectar from almond flowers stimulate the production of a great deal of worker and drone brood, and this must be sustained by the new locations because almond fields become an ecological desert after bloom.

Early in the season, all colonies are fed, often with a two-gallon food can in the lid of each production colony. Protein patties are also used to help the colonies build in bee population as necessary.

4. At the end of the shaking season, the colonies are moved into honey production areas. In Olivarez's case, the bees are moved to

Shaking screen cage and bulk box below. The top box of brood is placed over the screen and each frame shaken. The bees fall and the worker bees to crawl through the wires.

Montana where they produce a summer crop of honey. Natural food from pesticide-free areas is best for keeping colonies vigorous and able to repeat the queen package bee production cycle the next season.

SHAKING BEES

The shaking process is intense beekeeping, ending up with tens of thousands of bees in the air and landing on anything they can find. (don't worry, they find their way home). This can be intimidating for inexperienced beekeepers. The bee crew are well-protected with bee-tight bee suits and veils. Shaking packages is hard and sweaty work. It requires attention to detail so bees are handled with care and not hurt in any manner.

There are abundant bees, both workers and drones. The worker bee population of these colonies have been built up through feeding over the winter and early spring and may have been in the almonds in California. Package bee producers do not want to ship too many drones to their customers. Customers do not want drones in their packages: they don't build the colony. Most important, the queen and package producer need them to mate with all the queens they are producing.

Using the shaking screen cage, a hive body containing brood frames is placed over the cage, and each frame shaken so bees fall into the excluder area.

Years ago, I had a simple shaking funnel made by a heating and cooling fabricator, large enough to hold one or two frames of brood (after locating the queen). I'd shake the frames so the bees fell into a shipping cage. But this method resulted in many drones in the package, because young drones are often found on the brood frames. We might miss a queen, since some colonies may have more than one queen in the spring.

The solution is to shake bees through queen excluder material into a bulk box that holds 20 to 30 pounds of bees at a time. Queens and drones rarely make it through the excluder wires.

Working one hive at a time, a crew member sets a hive body filled with brood from a strong colony on top of the shaking screen. Each frame is shaken inside the box, and the worker bees drop into the bottom of the screen. After all frames are shaken, the hive body is returned to the hive, minus their young workers. The bees are gently smoked to force most of them into the bulk box.

After several minutes, most of the young workers are in the screen box, but not the queen and drones. These can be carefully returned to their hive—where the bees settle down. Worker bees, who are often older field bees, may be flying in the air, but they, too, will settle down and return to the entrance of their hive. After an hour or so, the bees are usually back to foraging and newly emerged bees are recruited to feed hungry brood.

Bulk boxes contain the younger bees from several colonies. There is great confusion and the bees do not fight. These bulk boxes may be used to set up starter and finisher colonies early in the season and used to set up mating boxes once queen cells are ready for mating. As the package shipping season starts, the bees are measured out by weight in the apiary using a scale. Holding cages are shaken so the workers fall into cone-shaped funnel into package shipping containers. Most producers overfill the shipping containers to allow for dehydration that takes place during shipping.

STORAGE AND TRANSPORTATION

Once the packages are filled with bees and a feeder added, they are ready to be moved to a cool room where a mated queen is added.

With a hive body placed over the shaking cage, the beekeeper shakes each frame so bees fall and are stimulated to crawl through the queen excluder.

Bees are smoked lightly to move through the queen excluder wire on the bottom and sides of the shaking screen. The queen and drones are shaken back into the original hive and are not lost or put in the packages.

Once the bees are in the screen cage, they may be poured into a series of metal funnels and weighed. These bees are then poured into the shipping cages. A can of sugar syrup is added once the bees are in the shipping cage.

These shipping cages are linked together for handling. Air spaces separate the boxes so the bees are not overheated.

This queen has been harvested from a mating nuc by the queen yard work crew and carried to a central work area.

It may take more than one day to shake and prepare a large load of 1,000 packages or more. So as the packages are prepared, they are temporarily stored in a dark, cool room with abundant air circulation to keep the bees cool and quiet. It is from this room that the bees are loaded onto the trucks for transportation to their final destination.

Driving package bees to all parts of the country is a sensitive activity in more ways than just time. Package bee producers must carefully select drivers for cross-country delivery of the packages because not all truck drivers are trained in the transportation of bees over mountains, through deserts, through rain and sometimes snow storms!

In the Chapter Three we will look at the biology of queens and package bees. In Chapter Four we describe the installation of package colonies by the beekeeper who purchased the packages to install into their equipment.

—Chapter Three—

Package Biology

Queen Production

Producing queens to ship with package bees involves three steps:

1. Selecting a genetic stock that produces high-performing daughter queens.

2. Producing queen cells and resulting queens with high pheromone levels as evidenced by their attractiveness to workers.

3. Mating queens with a large, healthy and genetically diverse group of drones.

These steps involve standard beekeeping practices, but each requires considerable attention to the biology behind these activities. Though easy to overlook, understanding the biology behind packages is crucial to using package bees successfully.

Package shaking operation in Orland, California, near almonds, walnuts and in the middle of dairy country, providing food from trees, legumes and lupines.

In nature, colonies that actively produce new queens follow their own genetically programmed biological rules. Normal colonies are super-families comprised of worker bees produced from different fathers the mother queen mated with during her mating flight(s). Each of these drones contributes half of his DNA to his offspring, but only for his own worker daughters. While the other half of the workers' biology and behavior come from the queen, the composition of the resulting hive is very diverse because of the multiple drone fathers. This explains why sister queens that have mated at the same time in the same mating yard may produce colonies with very different behaviors and appearance.

This super-family structure has led beekeepers and bee geneticists to suspect that workers sharing the same father, called super-sisters, may work to make sure that one of their related young

Groups of four mini nucs for the production of queens. The wire hive stands keep the nuclei off the heat of the lava rock and away from predators.

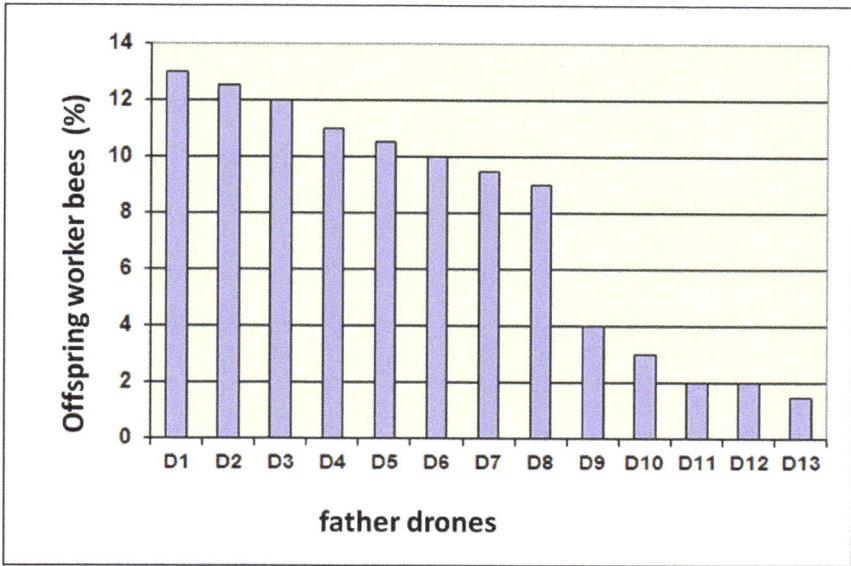

A queen's multiple mating results in a diversity of workers in the hive. This graph shows that the drones are not all represented equally. Workers that share the same drone father are called super-sisters. The number of drone fathers and sperm utilization affects genetic diversity within the hive. Koeniger.

worker larvae becomes the next queen, either during swarming or supersedure. Think of it as a struggle for power out of the plot of an HBO series. There is some indication that the dominant queen-drone sub-family may have some impact on the selection of the new queen, but super sisters do not seem to negatively impact the viability of the colony.

These behavioral coalitions, ruling families, political parties, mob groups—whatever you want to call them—influence many activities in the hive, but they seem to help the colony survive rather than cause harm. One family may carry unique genes that allow them to fight certain diseases. Another may have genes that favor the collection of a great deal of pollen. Another may precisely regulate in-hive temperature to help maintain homeostasis (environmental balance within the hive). The list of genetic possibilities goes on.

Beekeeper-breeders who specialize in producing queens often attempt to manage the genetic diversity of super-families by controlling the type or line of drones found within a mating area. This could mean the genetic improvement of thousands of colonies with higher vigor and tolerance of pests and disease.

These beekeepers seek one or more drone lines that possesses desirable genetics that eliminate negative traits such as defensive behavior, disease susceptibility and other recognized undesirable genetic traits. Selected queens that produce drone mothers (called drone mother breeders) produce queens whose drone lines carry desirable traits or behaviors that combine well with the queen lines.

Because these colonies produce worker bees that will fill packages, there may be hundreds or even thousands of these drone-mother colonies within flight distance of a mating area.

Over the years, many package bee producers have built a reputation on the desirable genetics of their bees, and for good reason. These beekeepers work hard to personally maintain the good stock, often focusing on honey gathering productivity, brood rearing, low defensive behavior and the ability to survive a year or longer without queen failure or colony death. They have fine-tuned the desired traits of bee strains and queen families until the performance of the final colony is somewhat predictable in the hands of the customer.

Another trait package bee producers may seek to replicate is some level of tolerance against the honey bee parasite, the varroa mite. Mite-biting and hygienic behaviors are examples of anti-mite defenses. Some producers know that their customers will use chemicals to treat for mites and other pests, so they do not focus

This queen is producing workers with different color patterns, perhaps reflecting different drone fathers and their superfamilies. T. Ives.

on mite tolerance or disease resistance. Other producers support those beekeepers who seek to minimize chemical use in the apiary, and actively select for tolerance mechanisms.

Some bee breeders take a more scientific approach to bee production by collecting data from a potential field of candidate queens that are being considered for the next generation of breeding stock. Beekeepers may measure brood to determine the colony's egg-laying rate, record the percentage of bees carrying varroa mite, or count the number of stings received while working a colony. Other beekeepers select breeder queens with a less rigorous method of data collection. They compare potential breeder queens from all of their production colonies and rank each queen's productivity in terms of honey production, gentleness, a traditional mixture of color, gentle temperament and vigor.

I know bee breeders who place push pins into the bottom board of a hive to indicate satisfaction with the colony based on what they see during colony inspection. By doing this each time they visit an apiary, they can quickly identify the colony that has the greatest number of push-pins and use these as grafting mothers. While some may consider this a primitive method, the basic premise has been used in the past.

Sampling for varroa mites using powdered sugar and half a cup of worker bees. New beekeepers should use this simple sampling method to keep colonies alive.

Worker bee biting a mite. Mite-biting is a heritable trait, passed from one generation to the next.

In the 1940s Dr. G. H. Bud Cale Jr. drew stars on the cover of the hives of potential colonies that would be used for inbreeding, leading to the name of the Starline hybrid.

GENETICALLY IMPROVED, INSTRUMENTAL INSEMINATED (II) BREEDER QUEENS

Other package bee producers receive breeder queens from professional bee producers such as Valerie Severson, queen breeder and owner of Strachan Apiaries, Inc. in Yuba City, CA. She instrumentally inseminates **New World Carniolan** queens.

Strachan Apiaries, Inc. is the largest woman-owned beekeeping operation in the United States where Severson produces 200 instrumentally inseminated queens received from breeder stock obtained from Sue Colby, a bee breeder formerly located in Davis, California but most recently working in Washington state. These breeder queens are instrumentally inseminated and produced in June of each season. They are used to provide grafting mothers for the following year's production season and to be sold to other beekeepers.

Valerie Severson and son operate Strachan Apiaries, producers of New World Carniolan queens.

Strachan cell producing area.

Brood pattern of a Saskatraz queen.

Several queen producers rely on this Colby-Severson breeder-queen-production system as a means of producing uniform New World Carniolan breeder queens.

Another program is the **Russian queen program** was started by the USDA Bee Lab in Baton Rouge, Louisiana and is now managed by a corporation consisting of owner-operator commercial beekeepers who are also bee breeders, each of whom maintains one or two queen lines. These lines are maintained by sharing virgins and drones on a modified closed-population system. (A technique explained in Laidlaw and Page's *Queen Rearing and Bee Breeding* and Connor's *Keeping Bees Alive*.)

The Saskatraz Honey Bee was developed by Meadow Ridge Enterprises, Ltd. working with the University of Saskatchewan and queen breeders from Manitoba and Saskatchewan. In the United States, daughter queens are produced by Ray and Tammy Olivarez (OHB, Inc.) in California using rigorously selected queens produced in Canada. Besides their wintering ability, this stock is selected for its excellent honey production, gentle temperament, increased varroa tolerance and resistance to brood diseases, as well as increased hygienic behavior.

Colonies in Russia tested by Dr. Tom Rinderer to select for use in the United States Russian bee-breeding program. USDA.

The Buckfast Bee program was developed by the late Brother Adam of Buckfast Abbey. Since his death, a group of dedicated bee breeders provides breeder queens to cooperating beekeeper-breeders including a group of dedicated amateurs located in Europe. The closest program to the United States is located in Ontario, Canada, where queens can be purchased from several queen producers. I have visited this program and consider it worthy of consideration by serious beekeepers. Because these queens and their colonies are productive and desirable, these bees are worth the extra effort and costs to obtain them.

The Minnesota Hygienic Stock (University of Minnesota), the **Varroa Sensitive Hygienic** (USDA Baton Rouge Lab) and other programs depend on selected beekeepers who use breeder queens to produce naturally mated production queens. I have used breeder queens from both programs and found them to be excellent stocks with reduced mite populations. Some beekeepers have had difficulty finding the production queens when they go to place orders in the winter. The solution is to estimate your queen needs early and place orders in the fall of the year before you need the queens.

The **Purdue Mite Biting Bees** carry a behavior of grooming that removes mites. The bees bite the mites, chew off their legs and antennae, make holes in their carapace (shell) and cause mite death. Unlike the hygienic trait, which apparently involves a number of recessive genetic traits, the grooming behavior seems to combine additive genes that encourage grooming behavior. This stock has been distributed to mid-western states, but their availability is limited.

Some package bee producers produce their stock in a Sun-Belt location and do not select bees where they are used in a northern state. These producers are often criticized for this practice. Fortunately, many end-user beekeepers obtain packages that contain queens from stock selected in northern locations and make them available as queens in packages. The Minnesota Hygienic, New World Carniolan, Saskatraz and Purdue Mite Biters are northern-bred queens. Keep in mind that where a queen is raised does not change the genetics of the resulting hive.

A few beekeepers, like Ted, Linda and David Miksa, sell ready-to-emerge queen cells from their Florida location from several stocks, some of which were selected for northern conditions. This provides beekeepers a relatively inexpensive method of obtaining improved

Grafting mothers or breeder queens in the Miksa operation in Florida.

Marked grafting mother for a breeding program.

genetics in their colonies. I strongly recommend that bee clubs place bulk orders for ripe queen cells for periodic shipment for distribution to members who will make mating nuclei and mate these queens to local drone populations. Do this well in advance of your expected needs. A few package bee producers will used these queen cells, mate them their apiary and ship them in package colonies, but only by prior arrangement.

GRAFTING MOTHERS AND GRAFTING

Breeder queens, or grafting mothers as they are often called, are usually maintained in a secure location, often locked up in a fenced area or even in open fence cages to prevent theft. The bees fly through the fence material, but humans cannot ordinarily access the valuable breeders. With some breeder queens selling for $500 to $1,000 each, security is an important issue for queen producers.

A small room where grafting takes place serves as a grafting room. There, you will find people dipping wax cells or assembling plastic grafting frames. These supply one or more grafters, people who transfer roughly 12-hour old larvae (after three days as an egg) from the bottom of a worker frame to a special cell cup. Royal jelly may be added to add nutrition and to increase humidity or the larva may be grafted 'dry'.

Frames of cells filled with the tiny larvae are carefully taken to a set of colonies called cell starters where they will be installed. These starter colonies are usually filled with thousands of young nurse bees, and they feed the larvae with royal jelly. Many designs do not have any brood present, and the nurse bees can be compared generally to un-milked cows, seeking a place to feed royal jelly and brood food. Several options exist. Some beekeepers used separate colonies for cell starters and cell finishers, while others use one type of hive that both starts and finishes cells. These methods are detailed in my book *Queen Rearing Essentials*. These details are not necessary or required reading for new beekeepers, but once you get into beekeeping for a year or more it is important to better understand how queens are produced.

The goal of this grafting process is to produce ripe queen cells. These are queen cells that have late stage pupae inside the cell that have nearly completed the third developmental stage of metamorphosis (pupation), and in a perfect world the young queen will emerge within 24 hours.

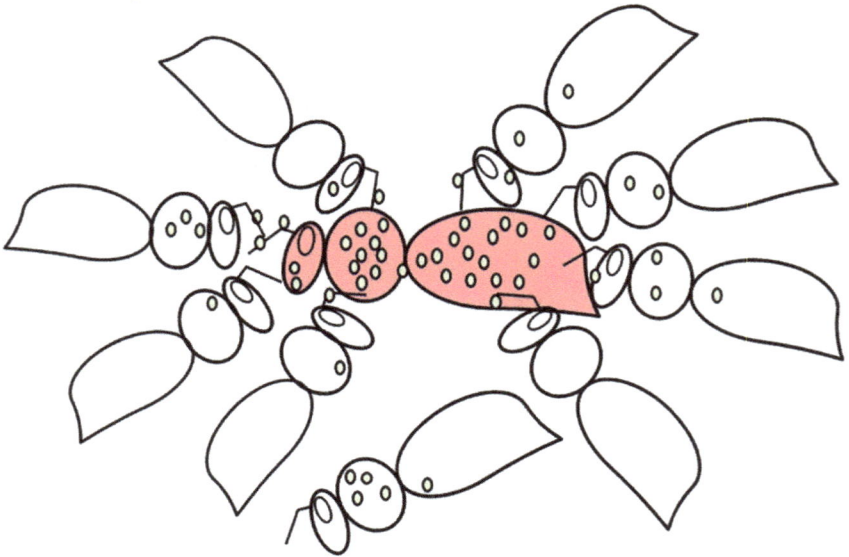

Representation of queen pheromone production and function: recognition, reproduction, egg-laying failing. Koeniger image.

Queen pheromone is a powerful attractant to drones, but only in the DCA. All these drones were attracted to this sex attractant locate inside this trap.

Population Management From a Queen's Perspective

As a result of evolutionary biology, a social insect like the honey bee has developed a system of chemical controls that are produced by the queen and passed from bee to bee. These chemicals, called pheromones, are produced by one member of a species and causes a behavioral change in another member of the same species. In a honey bee colony, queen bees produce queen pheromones, also called **queen substance**. Caron and Connor's *Honey Bee Biology and Beekeeping* delves deeper into the subject matter, so we will not repeat all that information here.

The key thing for beekeepers to remember is that queens produce pheromones throughout their adult life, but the amount and composition of the pheromones changes as the queen ages, mates, matures and produces eggs. It is generally thought that queen failure is associated with a reduction in pheromone production. This reduction may also be linked with the queen's egg-laying rate.

Queen pheromone production starts in the pupal stage inside the queen cell. At that time, the worker bees may remove some or all of the beeswax cap of the queen cell to expose the silk, thus allowing

57

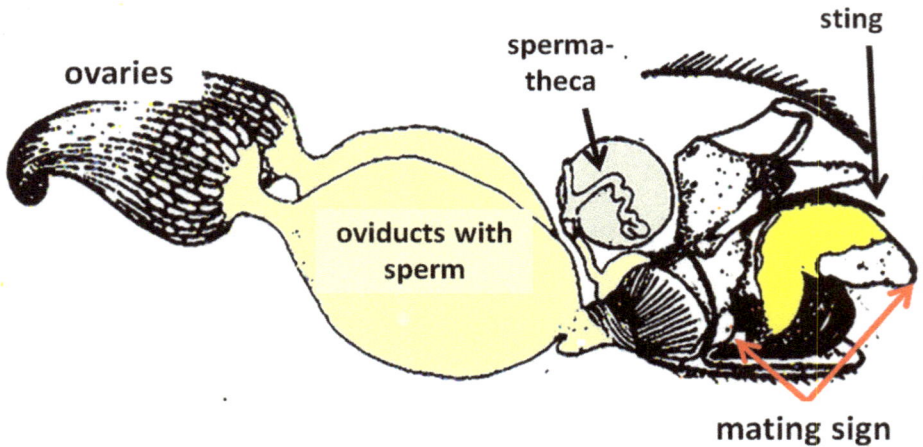

Queen's reproductive tract immediately following mating. The oviducts are filled with sperm and the ovaries are not yet swollen. The mating sign is in the sting chamber. Sperm must migrate into the spermatheca via the spermathecal duct. G. and N. Koeniger image.

the worker bees direct access to the queen's odors. Depending on the biology of the hive and the need for queens within that hive, the worker bees may allow the queen to emerge without interruption or intentionally confine a ready-to-emerge queen to her cell by resealing the wax so she cannot emerge. The bees feed the queen inside her cell but do not allow her freedom.

Once emerged, the queen's focuses on mating. She makes orientation flights, and at the eight to ten-day mark will make her first—and sometimes only—flight to a drone congregation area (DCA). This is an area she has never visited and may never return. There, she will mate with drones from area hives, but rarely her own, as a method to reduce inbreeding. If drone numbers are low or weather shortens the flight time, the queen will return to the DCA on another day. Queens apparently measure their multiple mating, each to a new drone, by the number of times they open their sting chamber and allow a drone to inseminate her.

Mating stimulates complex biochemical and physical changes in the body of the queen. Her brain produces hormones that trigger egg production, so her abdomen swells in both length and thickness. Eggs develop and travel down the medium oviduct to reach the spermathecal duct, where sperm are released and the

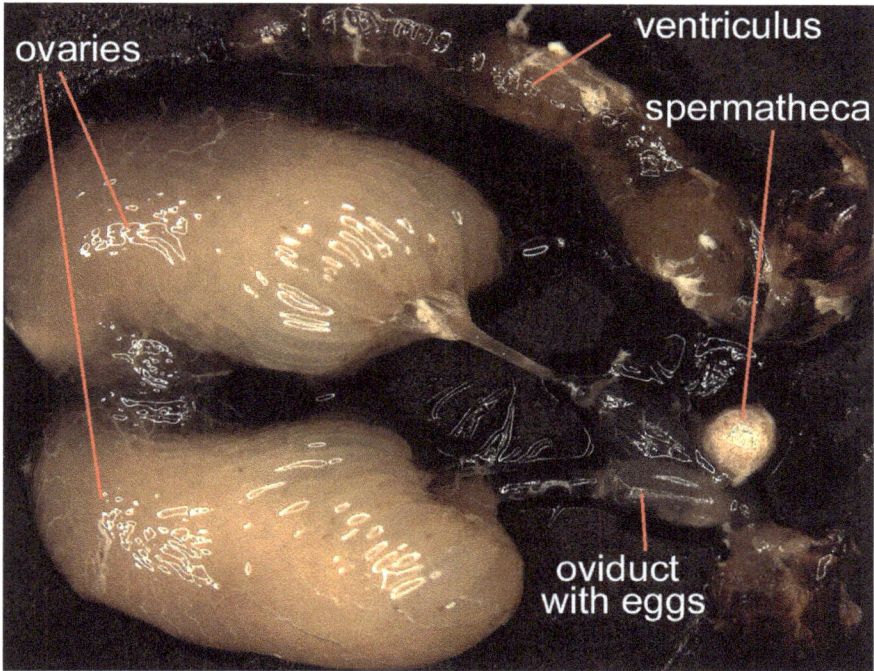
Ovaries in a laying queen, USDA.

egg is fertilized. Fertilized eggs have two sets of chromosomes, and the resulting bees are females (worker or queen). Unfertilized eggs result in males (drones).

Each queen may mate with 6 to 61 drones, filling her medium and lateral oviducts with the thick white fluid. There may be 80 million sperm in her body, but she only has the capacity to store between 5 and 8 million. The rest of the sperm are released and worker bees remove the thin threads of material after mating. This happens inside the hive. The queen will not fly again unless she leaves the hive with a swarm.

STIMULATING COLONY GROWTH

As described in Chapter Two, strong colonies are managed to eventually be shaken into packages. Top-quality queens must be installed in each colony and a combination of sugar syrup and protein patties fed when a natural food flow is not available. Bees must be fed well when they are in the larval stage and again in the young adult phase. This requires protein from bee bread and carbohydrates from nectar and honey.

Colonies are shaken two or three times. The first time is to set up mating boxes and queen production colonies. The second and third time are for production of packages for customers.

Some beekeepers use internal feeders (frame or division board feeders) or put holes in can lids and put feed cans into feed shells or over holes in the hive lid.

CONCLUSION

Understanding how queens work, how the mating process works, and how beekeepers produce queens will help the beekeeper understand how to use queens successfully. While based on simple biological principles, queen production and use is often complicated. This knowledge will hopefully ease the process of establishing packages.

The first critical feeding period of a queen is during the larval stage. These workers are building cells, secreting royal jelly and heating the cells.

Worker bee feeding a larva inside a queen cell on the face of a brood comb.

Once the egg hatches, workers continue to feed the larva until the cell is sealed.

Secondary feeding of a queen. Worker bees continually provide food for the queen as she lays about one egg per minute during peak season.

Mated queens are held in special holding frames like this one for several days to several months. The frames have brood next to them, to ensure care from nurse bees.

Metal cans filled with sugar syrup provide continuous feeding during queen production.

Queen bank holding six frames of queens prior to shipment in package colonies.

Queen candy provides food for the queen during shipping. Once a cork is removed, the workers eat the candy to free her from her wood and screen prison. Some beekeepers rush the process and soften the candy, which is not advised because the queen may emerge before she is fully accepted and may be killed.

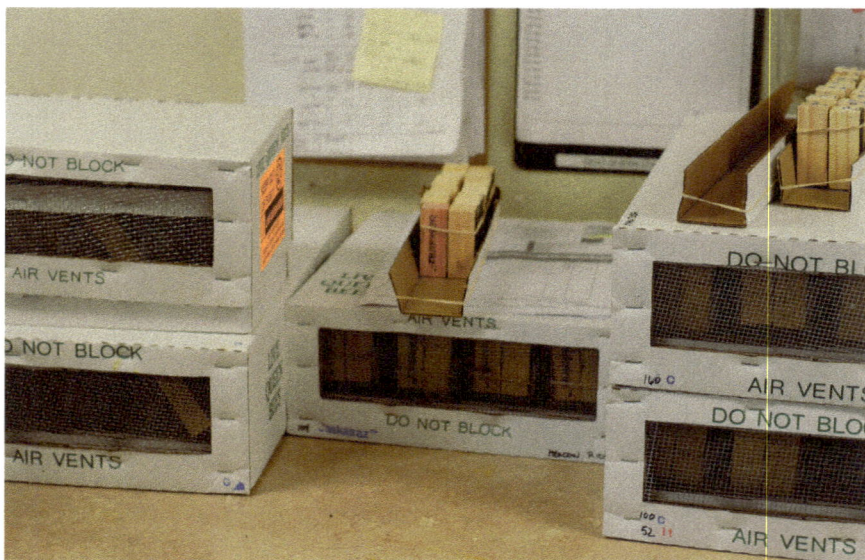

Small orders put workers into cages with the queen, while larger orders have the workers in a bulk cage.

Young nurse bees are shipped with large queen shipments. About a pound of young bees surround the queens.

Even in this Hawai'ian paradise, bees that produce queen cells must be fed. Here, sugar syrup is being poured into division board feeders.

Many queen and package shippers routinely treat along the ends of the brood frames with antibiotics to prevent American foulbrood from developing.

—Chapter Four—

Installing Packages

Two Methods To Introduce Packages

Like many beekeeping activities, there are multiple methods to reach the same goal. The objective of a beekeeper who purchases package bees is to install the bees and the queen into a hive equipped with either foundation/starter strips or drawn combs. In this chapter we will describe two methods you may review for use to install one or more packages of bees.

The first method uses the pheromones of the queen honey bee to draw the bees out of the shipping cage. This system is more conservative, allowing fewer bees into the air where they may drift away and be lost. It does, however, require a second visit to reposition the shipping cage and release the queen. In the

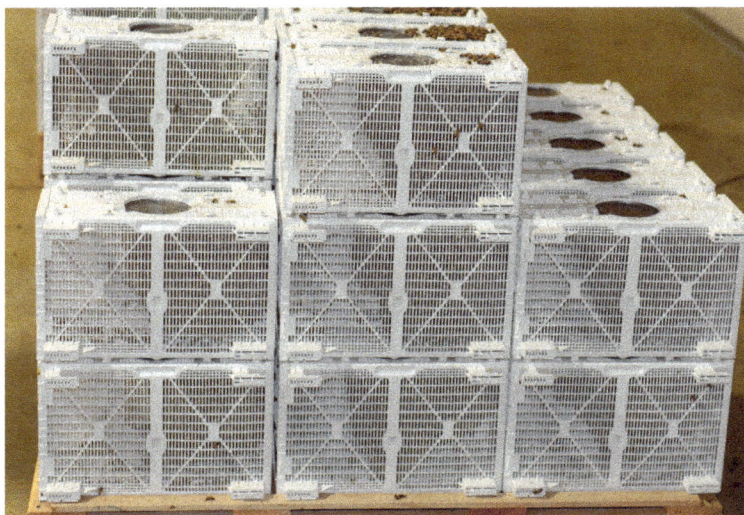

Packages of bees ready to ship to customers. These are plastic shipping containers used by OHB, Inc. in California.

second method, the beekeeper physically shakes the bees out of the shipping box and pours the bees into empty equipment. This system is fast but may not be suitable for use during a snowstorm. Because many bees are physically shaken out of the colony, some drift away, and either enter other colonies or are lost.

I show how these two systems work with two photo essays showing two kinds of shipping cages: wood and screen and plastic packages. The package producer decides which type to use, and the customer usually needs to accept the packages that the producer provides.

Newly installed package colonies must be fed with frames of honey and/or containers of sugar syrup. This is critical for rapid comb building, food storage, brood development, and colony growth.

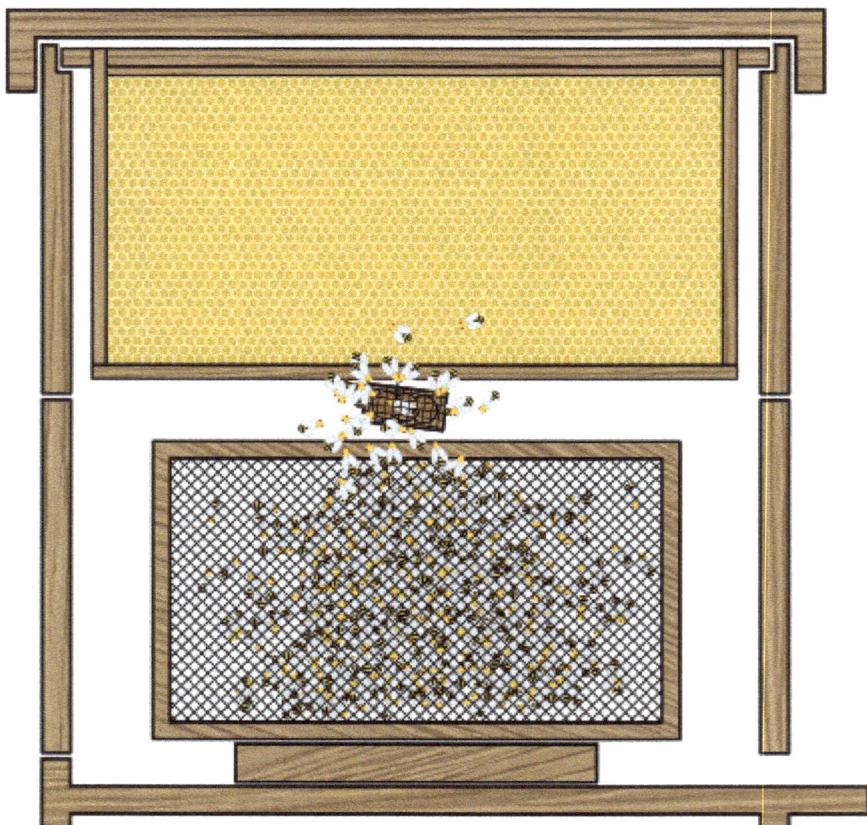

The drawing shows a simple method to remove bees from a cage by placing a queen (in her shipping cage) above the cage opening and using her pheromones to draw the bees around her. J. Zawislak image.

Day 1— Installing Package Into Equipment

The pheromone release system is pretty simple, but it requires some concentration. It works best with two hive bodies.

In the lower box, position the package of bees in the center of the box. Frames of foundation, drawn comb and honey should be on either side of the shipping cage.

Carefully remove the lid and extract the can of syrup and the queen cage containing the mated queen. Check to make sure the queen is alive and appears healthy. Set the can aside if it still contains syrup. Recycle the can if it is empty. Make sure the queen cage is not put into the sun which would cause the queen to overheat.

In the second (or top) box, fasten the queen cage with the live queen at the bottom of two combs so that when the top box is placed over the bottom box, the queen cage is positioned directly over the opening of the shipping cage. Fill the top box with frames of foundation and/or drawn comb, and if you have them, add two frames of sealed honey from an overwintered hive. This is in addition to sugar syrup you will used for feeding.

Let the pheromone system work overnight. During this time, the bees will crawl out of the cage to cover the queen and frames. The next day, reverse the two boxes: place the top box onto the bottom board and raise the bottom box above the first. As you do this, remove the emptied shipping cage and put it in front of the entrance to encourage any stragglers to enter the hive.

Carefully remove the queen cage from between the frames. Check to see the queen is alive. Then remove the cork or plastic cap from the candy end of the cage. That will depend upon which queen cage you get, a California cage, a 3-whole cage or a JZ's-BZ's cage. Remove the cork with your hive tool or a needle-nosed pliers.

This will allow the queen to be released from the cage once the bees consume the sugar candy plug. If the queen cage does not have a candy plug, you have two options. First, you can very carefully remove the cork and place the cage on the top of the frames and allow the queen to walk onto the frames. Mated queens often make a quick dash for the space between frames to avoid the light. They may also search for a drink of honey. This is called 'walking out the queen' and it is fun to watch. This allows the queen to start laying eggs in a matter of hours, and when you recheck the hive in three to seven days you should see eggs and larvae in the cells. The second option is to fill the cage with queen candy or marshmallow to delay the queen's release.

Removing the inner cover, a puff of light smoke on the top brood box will subdue the bees. A small bee population is not usually defensive.

This package was introduced the previous day. Bees are in both deep boxes.

Day 2— Removing Package and Releasing Queen

On the second day, reverse the hive bodies and release the queen. Honey frames or sugar syrup should be provided.

Removing the top box and the caged queen revealing the shipping cage beneath in the bottom box. Gently set the top box on the inverted outer cover so as not to crush the bees.

Setting the top box aside on the inverted cover at an angle. Most of the bees have exited the shipping package cage.

Carefully remove the shipping cage from bottom box. I do this bare-handed but first always look where I put my fingers!

Note the other shipping cage leaning against another hive's entrance. This process can be done multiple times in succession. We installed four packages in this series, so several hives are show. All are basically identical.

Place the empty shipping cage near the hive entrance. There are usually some bees in the cage, but after removing the cage from the hive and placing it at the entrance, the bees will be attracted to the queen pheromone and bees inside.

Bees lingering in and on the shipping cage. Many of these bees are exposing their scent glands. They will eventually crawl inside the hive.

This shows the space the shipping cage had occupied. Don't even think about leaving this space without frames or you will return to a mess of burr comb. Add frames to fill this area.

Fill space with drawn combs or foundation. This box will be on top when we finish this process.

Remove the frame from the top box that was set aside earlier where the queen cage was installed on Day One. Remove the cage and return the frame to the open space.

The bees, having spent a day attracted to the queen's pheromones, have surrounded the cage.

Use a light puff of smoke to reveal the queen inside her cage. Carefully remove the cage with your fingers.

Checking queen in cage. Many of these bees are using there scent glands to attract their hive mates. It's the bee equivalent of 'We Found The Queen.'

This California cage does not have a candy plug. Look carefully at the cork. It is partially chewed by the worker bees. Be aware that the bees may sometimes release the queen into the package box. In a perfect world, the cork will keep the queen in her cage before she is chewed out by eager bees.

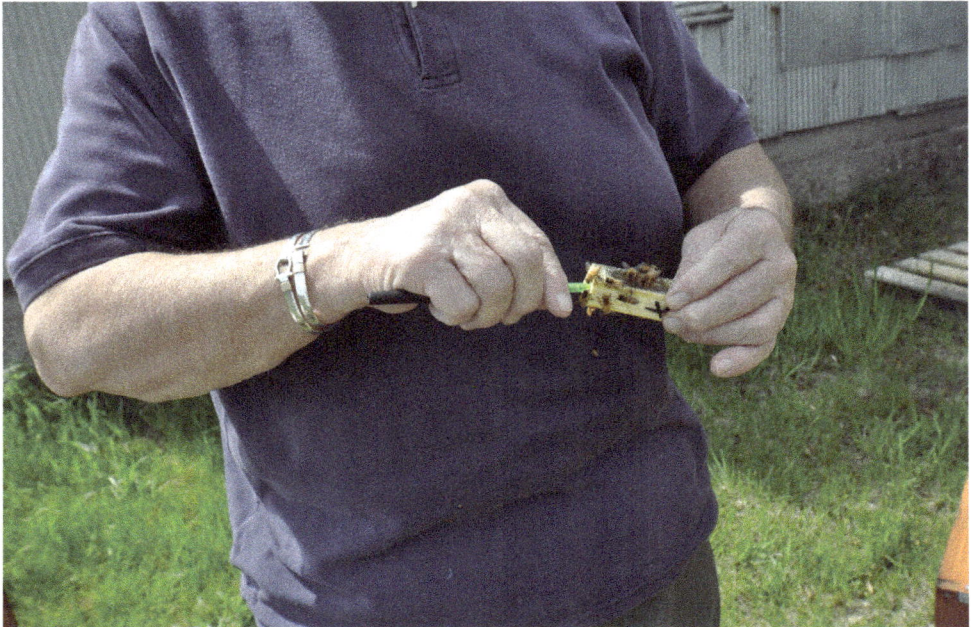

Because the cork was chewed so much, I used a pen to remove the remaining cork. Blocking the entrance with my finger to keep her from flying away, I placed the queen cage and queen on top of the frames.

Most of the time your hive tool will work to remove the cork.

Set the opened cage on the bottom box, allowing the queen to walk out. A puff of smoke may be needed to motivate a shy queen to exit.

Walking queen out onto comb. The queen walks out of the cage and quickly walks to the dark space between frames where she will soon start laying.

Bees liberated from the shipping cage immediately begin work establishing their nest.

Placing second box atop bottom hive body. The queen cage is placed on the front of the hive's bottom board. Residual queen pheromone may attract stragglers but also helps me to update my records on the queen type placed in each hive.

Newly released bees that have crawled up onto older comb.

Follow the photos for the second method: shaking bees.

This series of photos was taken at a June Field Day of the Connecticut Beekeepers Association when I lived in that state. The day was beautiful, and the bees were gentle. Better than installing packages in a snow storm in Michigan in March! Massachusetts beekeeper Vinny Gaglioni demonstrated package installation.

Here, we see a hive body filled with empty combs and foundation and the remaining frames. It is sitting on a bottom board, and the inner cover is about to be removed. We also have a package of bees (including worker bees, a queen in a queen cage, and sugar syrup), a spray bottle containing clean water, and a quart jar of sugar syrup (usually one part sugar and one part water).

An extra hive body and telescoping lid are on the far right.

Before opening the package, the beekeeper uses the spray bottle to mist the bees through the package screen to quiet the bees. This is very helpful if the packages have been allowed to overheat and have become excited.

The beekeeper will then open the package of bees and remove the syrup can and the queen cage, placing the lid back on the package until the bees are ready to be transitioned.

The beekeeper inspects the queen cage to make sure the queen is alive and has not been damaged. If the queen is dead in the cage, continue to install the package without her, but contact the package supplier for a replacement.

Finding that the queen is alive, the beekeeper removes the cork covering the end of her cage that has a wall of candy, allowing the bees to chew out this barrier once they are introduced. When sure the queen is fine, the beekeeper will place the queen inside her cage on top of the empty frames.

The beekeeper removes the lid of the package bees, then taps and rolls the bees out of the shipping cage and onto the surface of the queen cage, allowing the bees to adjust to the new queen and spread out.

Not all the bees must be removed from the package. By setting the package beside the new hive, the bees will be lured by the queen's odor and the scenting bees producing Nasonov pheromone.

The beekeeper gently pushes some into the open space in the hive body without disturbing the queen.

If left with the open space, the new bees may begin to build burr comb, resulting in wasted effort and wasted wax, so the beekeeper gently installs the remaining frames into the hive. Carefully space the frames to respect the bee space.

As the bees adjust to their new queen and home, the beekeeper carefully inverts a container of sugar syrup to provide food for the bees. Other feeder types may be used.

The beekeeper will then stack the empty (frameless) shell of another hive body on top of the bottom box, confining the bees and sugar syrup.

Finally, the inner cover and enclosing lid are placed on top of the second box. The bees and queen should be left alone for a few days. Make sure the bees have a constant food supply within each hive. Pick up the packages later in the day.

Once emptied of bees, the shipping packages should be picked up and either returned to the supplier for a deposit or, depending on the material, be recycled. I use empty package containers to house butterfly and moth caterpillers to study their metamorphosis.

CONCLUSION

Either method works fine to install package bees. The result should be the same: Happy bees!

In the next chapter, we'll discuss what to do with the bees and how to manage them once they are installed.

—CHAPTER FIVE—

INITIAL INSPECTIONS

Once you have installed your package colony into a hive and either directly released the queen or removed the cork or plug so the queen is able to get out, she should begin laying eggs. You will need to check on her success by looking for eggs. A package should contain one mated queen in her own cage. She does not need to go out and mate—you have already paid for that mating. Your queen should start laying eggs almost immediately after introduction and start building your new colony. If you check on the queen in the first nine days or slightly less, you can avoid big problems from queen failure that unfold if you wait longer.

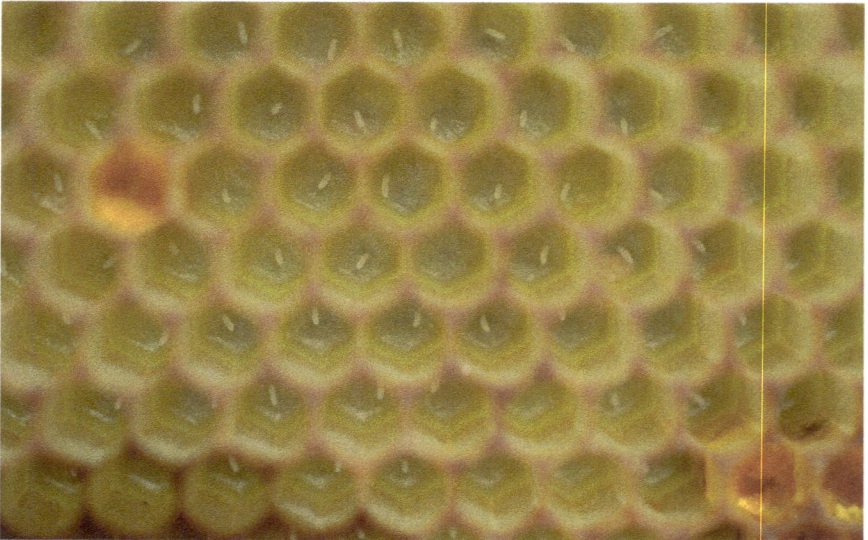

Newly laid eggs on fresh comb. The workers secreted beeswax to build the comb and the queen deposited eggs into the cells. In three days the egg shells soften and the larvae are evident on the bottom of the comb.

Search for Eggs

Honey bee eggs are tiny (1.5 mm) and positioned at the bottom of the cells. They resemble tiny rice grains. If the wax is new, the eggs can be very hard to see because they don't contrast much with lighter-colored wax. Use a small flashlight or LED lamp to search for the eggs in the cells. Expect to find eggs within the first 24 hours after introducing the queen into the bees, and larva between 4 and 5 days after introduction.

If you do not see any eggs 7 to 9 days after the package was introduced, you may want to ask an experienced beekeeper or your mentor to confirm that there are no eggs in your hive. You may have one of several problems. First, the queen may be absent or dead. This sometimes happens during the package introduction process. Second, there may be an un-mated queen in the colony shaken with the workers when the package colonies were produced. Soft-bodied young virgin queens may slip through the excluder. Look for empty, polished cells in the center of the combs. They indicate that a newly mated queen is present and is about to start laying eggs. These queens pose a problem when they are present because they will interfere with the introduction of a replacement queen.

A mixture of eggs and newly-hatched eggs. This comb is older and has had brood in it before.

When you have a good relationship with the supplier of the packages, they will replace a missing or dead queen without question or hesitation. If you have a queen problem, let them know immediately. Get a new mated queen and introduce her over a four-day period into the hive.

If you need to replace a queen you have about two weeks to get one installed. After a two week period of queenlessness, the introduction of a new queen often ends in failure. By then the worker bees, who have been busy collecting pollen and nectar and building beeswax comb, are going to become too old to accept a new queen. Some of the workers will have started producing unfertilized eggs. As the bees in the package age, some bees have been dying every day, and your package bee population is dwindling.

If you wait a month or more to replace a missing queen, it is nearly impossible to introduce a queen into a group of old workers. After a full season you will have so few bees left that you have little choice but to let the bees die or combine the hive with a strong hive with a good queen. Use a single sheet of newspaper to combine the hives if there are still enough bees to make it worthwhile.

Boosting Using a Frame of Brood

If you have other bee colonies, you are in the position to install package colonies and boost the package colony by adding a frame of brood. This may be done either when you install the package and queen, or it can be added a few days later. Select a solid, disease-free frame filled with mostly sealed and emerging bees but with some eggs and larvae too. The new bees are accepted by the package colony and will help fill an age gap in the bee colony demographics.

I like to add a frame of sealed and emerging brood for a second reason. If the queen fails during the installation and introduction process, the bees will select a few newly hatched larvae from the frame of brood and convert them into queen cells. Sure, this is not the highly improved, genetically-superior queen you paid big money for. If you allow the bee-selected queen to develop, emerge, mate and start laying eggs in a queenless colony, you may just save the colony. You could replace this queen later or just watch and see how she does as the queen of this colony.

A mixture of honey cells and older brood.

Worker bees feeding larvae. The bees have started to cap the cells by adding wax.

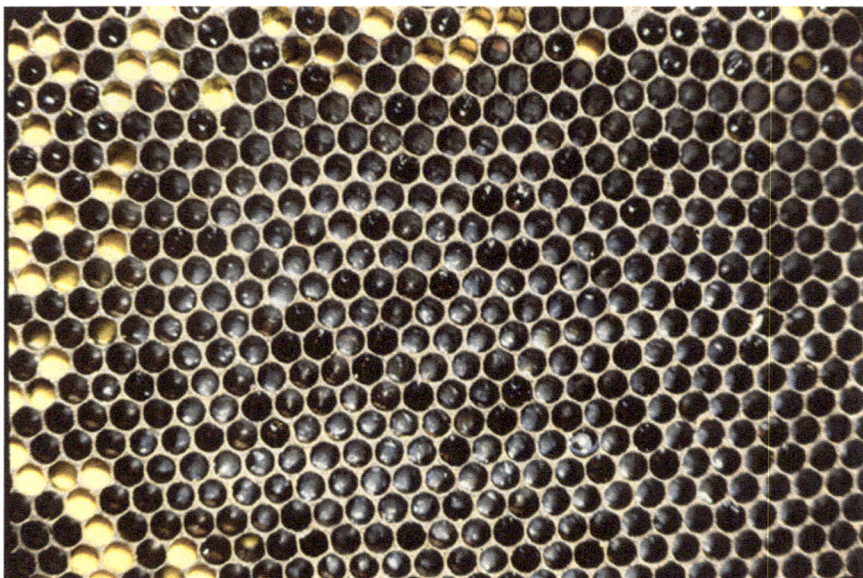
Two and three day old worker brood with cells of bee bread to the left and above.

Open brood. Here, the queen has had an earlier brood cycle as evidenced by the presence of a few sealed worker cells.

Seeing and Evaluating Sealed Brood

The development of bees, or metamorphosis, follows a set schedule. All eggs are present for three days. All larvae take six days in what we call the feeding stage, and all worker bees are in the sealed brood stage for a total of twelve days. Total development takes 21 days for worker honey bees of European origin.

Nine days is the earliest you expect to see sealed brood in the European races of honey bees. That means that it will take your package one week and two days before there is even the possibility of seeing sealed brood. The sealing of brood occurs at the end of the fifth larval day, when the larval bee is large and well fed. The cappings consist of beeswax from the hive. Some may be new wax and some recycled wax from other parts of the brood nest. The cappings are not air tight, and allow gases to flow in an out of the cells.

Once a brood cell is sealed, it will take 12 additional days before the bee emerges, for a total of 21 days for full metamorphosis. Throughout the 21 day period, the queen will continue to produce eggs and more and more brood will be sealed. It is a continuous process, and if a queen lays 1,000 eggs per day for 12 days, you should have 12,000 cells that are sealed (minus natural or disease losses). Since bees rarely fill every cell, especially in the corners,

Frame of about 80% sealed brood.

Sealed worker brood and some cells of pollen.

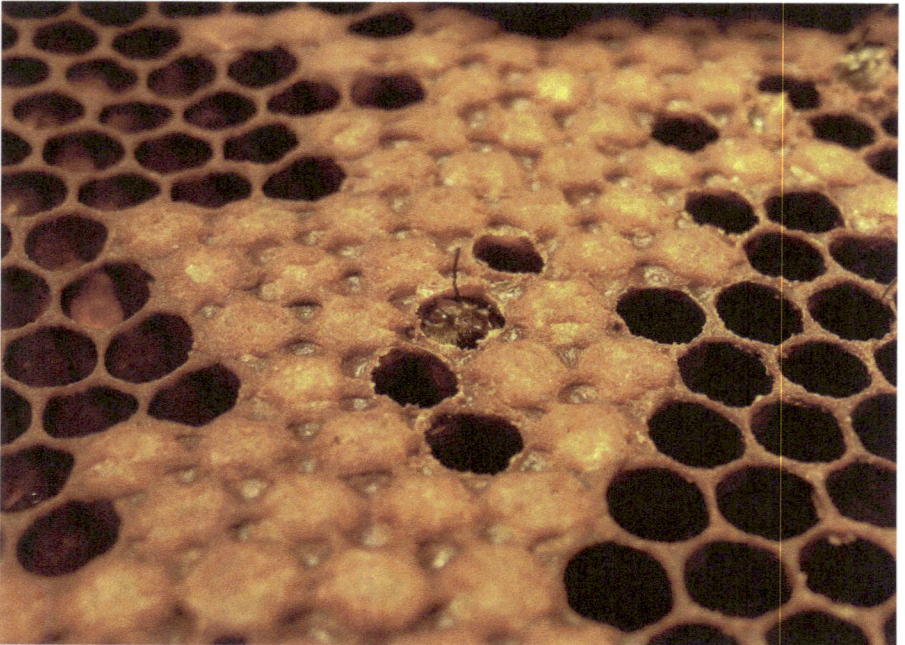

Worker bee emerging.

this means that there will likely be three to six frames with brood on it—sealed brood. Or, if the queen lays 1,500 eggs per day, then the colony should have 18,000 sealed brood cells, resulting in five to nine frames with brood on it. This will not be likely for most three pound packages, so do not be surprised if there is less than this, but all brood patterns should be relatively tight and healthy.

No Brood

If there are no eggs and larvae by the nine day mark after installing a queen, contact your supplier and request a replacement queen. Ideally, the person who supplied the packages is local and has additional queens in stock that they can give you for queen replacement. Ethical queen producers replace these queens without question or debate, but they may ask you questions to see if you have overlooked anything. Be honest to determine if you did something that killed your queen.

THE FIRST MONTH

Emergence of Workers

Three weeks after the queen has started egg laying, expect to find areas of worker brood that are emerging. The queen will deposit

Cells with multiple eggs, typical of laying workers. Workers eventually remove extra larvae and only one will become a worker-sized drone in each cell.

Open and sealed drone brood in worker cells laid by worker bees in the absence of the queen. Bullet-shaped drone brood is a classic trait of laying workers.

new eggs into these cells within a few hours and continue the colony buildup. Drone brood, if present, will take 24 days to emerge.

Multiple Eggs and Bullet-Shaped Brood

If a queen fails, and when there is no worker brood present, the worker bees are no longer inhibited by queen and brood pheromones. In just a few days the queen-less, brood-less worker bees will begin to lay eggs without the presence of pheromones to stop egg development in their ovaries. Hundreds of worker bees will lay a few eggs per day. Unfertilized, they develop into worker-sized drones. New beekeepers frequently fail to detect these drone eggs and larvae until it is too late to correct the situation. The best approach, in my opinion, is to combine a colony of laying workers with a strong colony containing a healthy, vigorous queen and lots of worker brood in development. Both queen and the brood pheromones will inhibit the laying workers from performing this wasteful activity.

The eggs from laying workers may be on the side of the cells because the workers' abdomens are sometimes shorter than the queen's and cannot reach the bottom. In other colonies where the cells are shorter, the eggs will be on the bottom of the cells. Workers do not recognize these eggs and larvae, and there are often multiple eggs in the cells. One drone is ultimately produced in each cell, with multiple larvae removed.

With queen cells, most of the time the cells are at the edge of the broodnest (L), but sometimes there may be just a queen cell or two on the face of the comb where larvae are selected for queen replacement (R).

With emerged queen cells, one cell should have a flap present where the queen cut herself out. Both of these cells emerged and the capping flap has been chewed awary by the bees. Eventually thse cells are completely removed, so you do not have much time to examine them.

As the founding worker bees age and die, the colony will dwindle, leaving a small colony of drones that cannot support or sustain itself. Some beekeepers have some success reclaiming these drone-laying colonies by adding frames of brood to the hive and letting the bees sort this mess out. To me, this is basically the same as combining the package hive with an existing nucleus or support hive, which is often the best solution to the problem. It calls to question the strength of the combined unit. I vote for a strong nucleus rather than a single frame of worker brood.

PREMATURE QUEEN REPLACEMENT

Unfortunately, it is not unusual for new package colonies to produce queen cells within the first few weeks of installation. The bees used eggs and larvae produced by the new queen. She may have produced one or more cycles of brood before replacement cells appear. Apparently, the worker bees evaluate the queen's performance and respond by producing a replacement queen. This may be a function of the queen's egg or pheromone production. Or it may be the bees' response to the small hive size found in package colonies, something the queen does not control.

There are many factors that contribute to this early replacement. The bees are apparently not responding favorably to the performance of the queen that traveled with them in the package and have started to replace her. This means that there will be queen cells in production in the hive within a few weeks or months after the package was installed. If the colony has been successful in producing worker brood, replacement cells will appear on the face of the frames. Most of the time the cells are at the edge of the broodnest, but sometimes there may be just a queen cell or two on the face of the comb where larvae were selected for queen cell production and replacement.

Early queen replacement often occurs when the queen is not producing enough eggs or adequate queen pheromone, or she has been damaged in some way (exposure to cold, pesticides and poor rearing conditions). I doubt that varroa mite feeding is much of a factor, but the viruses varroa promote may be an unappreciated contributor to queen replacement.

Don't panic if the bees attempt to replace the queen by producing queen cells. Do not destroy the cells. That might eliminate the possibility of the bees ever replacing their queen and may kill the colony! If you find sealed queen cells (the cell tips are closed), mark your calendar to recheck the hive in one week to see if the queen has emerged normally. To find the frame with the cell, mark the frame with a push-pin or a felt pen. In seven days, look at the location of the queen cell. The cell should still be present, and the tip should have a flap present where the queen cut herself out. If

Position the cage in the hive so the workers can feed the queen through the cage. Ideally, the queen should be with the bees for at least four days to allow her pheromones to mix with the bees and for her body to swell with eggs. Then remove the cork or, in this case, the plastic cap, so the worker bees are able to consume the candy plug and release the queen.

The queen is held in a queen cage during the introduction period. Workers are attracted to her pheromones and feed her through the cage.

there are multiple cells, some of them will have holes in the side where a sister queen chewed a hole and stung the queen inside.

Replacing a Queen with a Delayed Introduction

You have the option of getting a replacement queen and installing her. Hold the cage in the hive for four days so the workers can feed the queen through the cage. This is a delayed release introduction. Ideally, the queen should be with the bees to allow her pheromones to mix with them and for her body to swell with eggs.

Some 'experts' advise beekeepers to direct release the queen into a queenless colony, arguing that the bees are so 'hungry for a queen' that they will accept 'any' mated queen without failure. I strongly disagree with this approach. Yes, queenless workers are searching for queen pheromone, but they benefit by having the new primary reproductive introduced gradually or she will be killed in the process.

Keep in mind that the queen's interaction with workers is based on a highly evolved, and beautiful and annoyingly complex system of stimuli and responses. A queen's pheromone production is influenced by many factors: her genetic background, her age, the number and quantity of sperm she carries in her spermatheca, the

An entire slab of fresh beeswax filled with eggs and open larvae. What happened? A queen cage was placed between two frames to introduce the queen. The spacing stimulated comb construction. The comb may be installed into an empty frame and saved. This is a big investment by a new colony!

Burr comb built on foundation by a package colony, in part because the frames were improperly spaced.

number of drones with which she has mated, and undoubtedly factors we do not yet understand.

BURR COMB CONSTRUCTION

Any group of bees will produce some pretty wild comb. Some artists have used bees to create the artwork they generate. Kalamazoo Bee Club member Lad Hanka, has put his artwork into hives, perhaps inspired by some bee-generated work of art on display at the Museum of Modern Art in New York City. Beekeepers do not want unusual comb building in their hives as a general rule. They must learn proper frame spacing by respecting the bee space.

Cut the burr comb out of the hive and let the bees build the come with proper comb spacing. Large sheets or pieces of comb may be installed into empty frames, or hung by wire, string or plastic ties from top bars. It is a huge waste to discard large areas of sealed brood, as the bees have invested a great deal of energy in producing the comb, gathering food, and feeding the developing bees.

Kalamazoo artist and beekeeper Lad Hanka took a drawing he did for his father's mead bottles and incorporated the artwork into comb placed into hives for the bees to add their contribution. L. Hanka.

DRONES IN NEW PACKAGE COLONIES

Earlier I explained how most package bee producers remove workers from hives using screen cages that use queen excluder wire to keep drones and queens from entering the bulk bees used to fill package bee containers. Humans are often surprised to see how early new colonies of bees produce drone brood. Why, we think, would a new colony need to produce bees. Putting aside sexist explanations I have endured about worker bees 'needing' drones in the hive, I see drone production as a response to normal

Drone brood at the bottom of a frame of worker brood and stored honey.

Full sized adult drone emerging from dark comb at the endge of a frame (L). Compare a worker bee (11 o'clock), a full-sized drone (5 o'clock and a dimunitive drone produced in a worker cell (8 o'clock). A worker-sized drone is emerging from the broodcomb (R).

population growth and one aspect of a healthy colony, where drone production is routine and part of the colony's evolution-driven behavior.

New white comb and fresh nectar/honey can be produced by a good package.

Sealed brood and cells of bee bread, ready for the next brood cycle.

New package colonies quickly build comb, collect and store pollen and convert it into a fermented bee bread, and collect nectar and store it as honey.

FEEDERS

There are many types of bee feeders: glass jars and jugs, metal pails (new metal paint cans), plastic 'frame' feeders that replace a frame of comb in the broodnest, and entrance feeders. I have seen feeding jars that fit into a hole in a single-board lid. Western commercial beekeepers use metal 'paint-thinner' cans using top lid openings. My favorite method of feeding a colony is still the addition of frames of sealed honey to a colony, the number determined by colony need. Protein and sugar feeder patties are beneficial in the late winter an late summer if needed.

Jar feeders: Large jar on package in Connecticut (UL); Entrance feeder on new package in Michigan, (UR): Jar feeders on new nuclei in Florida (LL), and Jar feeder on mating nucleus in Texas (LR).

Metal cans feeding cell builders in California.

Frame or division-board feeders. Plastic (L) and wood (R).

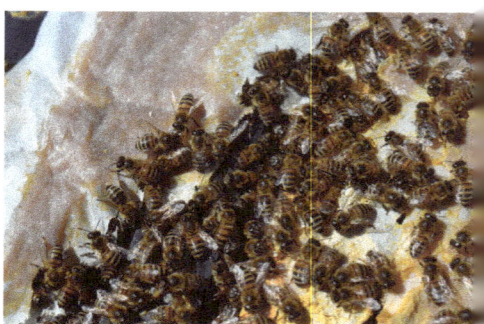

A lovely frame of honey from my colony in Michigan (L). Winter feeding pollen patty which is over 90% sugar (R).

—Chapter Six—

Summer, Fall, Winter

Meeting Goals in Summer

Once your package bees are installed inside the hive and the queen is laying hundreds of eggs every day, what is your job during the next few months?

Review your goals or expectations. Most new package beekeepers install their bees in new equipment with sheets of foundation or starter strips. Repeat package users may have used packages to restock colonies that died over the winter, and are busy expanding their operation using combs that were drawn out the previous summer. They may also have full frames of honey to use to boost the bees.

If you install a package bee colony around May 1, give yourself three months (90-100 days) to let the colony build. By early August, the colony could be basically identical to any previously overwintered colony in the apiary. If you are a new beekeeper, visit an established beekeeper to see what their overwintered colonies look like. Look for two or three deep hive bodies or their equivalent and several boxes of stored honey.

Mid-August is when the worker bees being produced are often the ones that produce the bees that enter the winter months. These August bees must be well fed if they are to produce well-fed Winter Bees or Fat Bees. By making sure the August bees are very well fed—something that is often difficult because of a nectar and pollen dearth—beekeepers help the colony survive the winter months. After a very strong nectar flow from mid spring and to some point in the summer, colonies need abundant floral sources or you must provide backup feeding if there is a gap in the natural food supply. Your goal is to support the queen so she can produce as many eggs *every day* that she possibly can, and protect the bees that will either cause bee losses, like exposure to insecticides, or prevent them from building in numbers, like a drought.

In most cases, a new package installed on foundation or given starter strips probably will not develop a surplus of honey during its first year. While we sometimes hear of 100 or more pounds of honey produced by package colonies, we may need to wait another ten to twenty years for this to happen again!

Most beekeepers want their bees to fill the equivalent of three deep boxes with beeswax, bees, pollen, and sealed honey. Good colonies will do that, but it is a great deal of work for the bees. It takes bees eight or more pounds of honey to produce one pound of beeswax. So, to build three deep boxes of beeswax comb, your bees need to collect, process and consume a great deal of nectar.

One option is to reuse frames of comb. Drawn comb frames must be inspected and determined to be free of American foulbrood and selected from colonies with no history of heavy virus infestations. By doing this you will provide the bees a great boost. The bees will immediately clean the cells and use them to store nectar and pollen or to raise brood. When this happens, more of the hive's resources are put into the production of a large population of well-fed bees so more honey will be stored for the winter.

If there is a history of poor nectar production in your area during part of the spring or summer, plan on feeding sugar syrup to produce beeswax to compensate. Beeswax produced from sugar is chemically identical to beeswax produced from honey (but lack the pigments found in nectar). From the day you install your bees, feed them. Keep the sugar syrup on the colony. Do not stop until the bees are working on the equivalent of a fourth deep hive body. That could be all summer, depending on the natural food intake.

Once you have drawn comb and maybe extra frames of honey, you can effectively use package colonies to full advantage.

POPULATION GROWTH

A three pound package contains 10,500 bees (3,500 bees per pound x 3 pounds). A 3-deep colony contains 45,000 to 55,000 bees. That is a 4.5 to 5.5 x increase. During the early period of a package colony's growth, a large percentage of the older bees focus on brood rearing rather than honey storage. In small colonies, like a package hive, the bees concentrate on brood rearing and population growth. Only when colonies become very large do the workers shift more to food gathering and processing.

Brood rearing reduces worker bee longevity. Brood rearing and the presence of brood in a hive shortens the life of bees, explaining why worker bees produced in the later summer and fall live longer than worker bees produced in the late winter and spring.[1] As a colony grows in population it will dedicate a larger percentage of its workers to surplus food gathering and storage.

PREPARING PACKAGES FOR WINTER

Queen Status

As colonies start preparations for winter, a colony's success in wintering may be significantly impacted by queen replacement through supersedure. Summer queen replacement is not uncommon, as it is a natural time for colonies to replace queens, even queens in packages, swarms and increase nuclei.

We do not expect package colonies to replace a large number of queens, but perhaps as many as one third of the queens will be replaced. Since this is a natural biological process, the beekeeper must decide if they must intercede and introduce a new queen.

---

[1]"We demonstrate that brood rearing reduces worker vitellogenin stores and colony long-term survival. Yet also, we establish that the effects can result solely from exposure to brood pheromone. These findings indicate that molecular systems of extreme lifespan regulation are integrated with the sensory system of honeybees to respond to variation in a primer pheromone secreted from larvae.

"We show that changes in storage protein physiology, which can affect worker survival directly (Koehler, 1921; Maurizio, 1950; Maurizio, 1954; Haunerland, 1996; Seehuus et al., 2006), are influenced by a primer pheromone secreted by larval brood. This finding calls for a revised understanding of how diutinus bees develop and of how successful overwintering is achieved by honeybee colonies.

"It has already been proposed that brood rearing shortens worker life and can lead to colony deaths in winter (Eischen et al., 1984; Omholt, 1988; Fluri and Imdorf, 1989; Amdam and Omholt, 2002) but the cause–effect relationship was previously explained by the metabolic costs of caregiving (Amdam et al., 2009)."

B. Smedal, M. Brynem, C. D. Kreibich and G. V. Amdam, Brood pheromone suppresses physiology of extreme longevity in honeybees (Apis mellifera). The Journal of Experimental Biology 212, 3795-3801

The option of letting the bees replace a queen at any time carries with it the risk of failure in this process. About one quarter of the queens are lost due to orientation problems, predators and the weather. These cause the replacement queens to fail to return to their hive and start laying eggs.

If you have a nucleus support colonies in a thriving apiary you already have the best way to solve the problem of a summer queen replacement. Young laying queens in backup nuclei are a quick and easy way to renew a queenless colony caused by mating failure.

Stores

Many beekeepers with new package colonies are well-advised to leave the honey on the colony for the winter. Any surplus honey can be removed in the spring or used to build up new colonies, especially new package hives. There may be favorable economics supporting the removal of honey for resale and feeding back lower cost sugar syrup. This idea appeals to some beekeepers, but requires more work to achieve, so often the simple approach of leaving the honey on the hive is an acceptable practice.

Mites

Tracheal mites do not seem too much of a problem for most North American beekeepers, but the varroa mite is still a huge problem. When combined with the issues caused by various bee viruses, varroa mite levels must be monitored and managed.

Protection

Colonies may need protection from high winds, cold temperatures and other risks. If colonies have excellent wind protection they may not require wrapping with roofing paper or insulation material. Some beekeepers argue that intense wrapping eliminates the need for an upper entrance. Several colonies may be grouped together and wrapped with a heavy insulating blanket. This appears to reduce the need for the bees to enter into a tight cluster during cold weather and allow the bees to reach more stored food. When the colony has a large population of worker bees and abundant stored food, wrapping is not a bad idea where winter is prolonged and steady. In regions where winter temperatures are up and down, colonies may not need to be wrapped at all. In these, hive top insulation may be the protection the colony requires.

Sampling a group of colonies or a sub-sample of hives in a larger apiary gives the beekeeper an excellent mechanism to observe mite numbers and track their development. There may be relatively few mites per 100 bees during the spring and early summer, only to have the numbers grow rapidly in the following season. Varroa populations may peak during the late summer and fall as surrounding colonies die and the mites migrate to healthy colonies. Many colonies that die in the late fall and early winter (before January) die because of declining bee numbers and increasing varroa mite numbers.

Varroa destructor numbers are best sampled by using one of several methods. One method accepted as the most sustainable technique because it does not kill the bees is the powdered sugar shake: Take 300 young adult bees (half a cup) from the brood nest and vigorously shake them with 2 or 3 tablespoons of powdered sugar for one minute to dislodge the mite from the body of the bee. This method does not kill the bees but dislodges about 90% of the mites from their bodies. The mites and sugar can then be shaken out of the jar and the mites counted. A mist of water will help dissolve the sugar and make the job of finding and counting the mites much easier. The bees are then returned to their hive where their hive mates clean off the powdered sugar.

Following a late summer or fall sampling using the sugar shake, if the test produces more than two mites per 100 nurse bees, the beekeeper should seriously consider using one of the various treatments, starting with the natural compounds and organic acids and oils. These compounds are effective at different temperatures and hive conditions. Consult with the Honey Bee Health Coalition for details.

LOSS OF A PACKAGE COLONY

Reasons for Loss

There are many reasons for colony losses during the summer and fall. A well-managed apiary has losses below 20% annually.

Starvation

The lack of adequate resources in terms of sugars or honey often lead to colony death. This may happen at any time during the winter, but in northern areas the problem is often seen in March and early April when colonies are in their rapid growth period and stored food supplies are dwindling quickly. Feeding with sugar blocks, dry sugar and with frames of honey are recommended. Feeding with other materials containing high levels of waste products, may lead to dysentery and colony loss.

Poor Summer Protein Levels

Low protein supply during mid to late summer often results in losses the following winter. The reason lies in the multilayered aspects of bee colony development.

August in many locations is a time of low protein intake because there are fewer flowers in bloom. Ironically, efforts to remove 'invasive' plants like purple loosestrife have eliminated major pollen sources blooming during the mid to late summer. If the bees fail to obtain a good pollen flow from late summer, flowers like goldenrod, asters, heartsease, and smartweed, there will be low protein nutrition in the bees produced at that time.

These bees will produce the last of the bees of the season. If they are nutritionally weak and carry low protein levels in their bodies, the bees going into winter will be poorly nourished and will die quickly as they start brood rearing in the winter.

Pesticides kill bees directly and indirectly. Several compounds do not have warning labels to protect honey bees, but, when combined with other compounds, they create a synergistic impact on the bees. They may not forage or reproduce normally. Herbicides specifically reduce the abundance of bee forage and negatively impact the nutrition of thousands of colonies.

High Varroa Mite Load

Colonies with more than two mites per 100 worker bees will be affected by these mite levels. The bees will be weakened as a result of the feeding by them mites, leading to an early death. A pre-existing virus load will combine with other damage to cause bee losses in the late fall and early winter.

Small Colonies and Very Low Temperatures

Since the varroa mites have become established, many beekeepers see colonies dwindle to small numbers. In the late winter and early spring, they may have a few hundred worker bees, a viable queen, and several square inches of brood. Because the colony is small, it does not cover a large area on the comb, and because brood cannot be abandoned, the bees will not move to available stores in the hive. Any cold snap will literally freeze the bees in place.

Queen Replacement Failure

In the summer and late fall, any queen replacement effort from either swarming or supersedure is associated with queen replacement failure in one out of four colonies that share this experience. Because the colonies may be strong and have boxes of stored honey, this queen replacement may not be noticed by the beekeeper. Queenless colonies become populated with laying workers. These bees produce worker-sized drones, and the colony collapses due to this hopeless situation.

Beekeeper Errors

Beekeepers cause many problems when they use package colonies. In the next chapter we will discuss many of the issues we hear who are new to beekeeping as well as a few from more experienced beekeepers.

Frame with abundant stored pollen, processed into bee bread (brood food).

—Chapter Seven—

Problems with Packages

Timing is everything. Package bees that are installed too early or too late in the season pose unique challenges and problems to the inexperienced beekeeper.

Some packages become available quite early in the season. One example: receiving packages in late March in New England. Think about that if you live there. Quite often there is a big snow storm in late March or early April. If offered, the inexperienced beekeeper should pass on such an early date. An experienced beekeeper might take the chance if, 1. They have frames of drawn comb, sealed honey and pollen that can be used to set up the colonies, and 2. They are prepared to feed the colonies with sugar syrup and protein patties for months or even all season.

I have watched beekeepers install packages by shaking bees into hives during a snow storm. It probably does nothing to help the bees. Some bees are lost as they hit the snow and are chilled. Use the pheromone release introduction system to reduce stress on the bees. Don't mist or wet the bees during installation.

The opposite extreme is the receipt of packages in mid June. In most areas, the nectar flow is underway and it would be July before the first cycle of brood emerges. These are the bees that will gather nectar and pollen in large volume for the season. The inexperienced beekeeper should pass on such late colonies. Only experienced beekeepers with existing colonies should consider obtaining them. Then they can make up a combined package and nucleus by adding one or two frames of emerging brood to boost the colony with young bees that will help the colony collect the all-essential food it will need to enter the winter.

The Queen is Dead

Queens can die during shipment. They must survive human removal from the mating nucleus, introduction into the package container, and travel on a truck for thousands of miles. Sometimes, queens die during this set of experiences.

California package producer OHB, Inc. recommends you leave the dead queen in the shipping cage and obtain a replacement. Once the new queen is received, take out the dead queen and hang the new queen in the package for 12 hours before introducing the package into your hive.

The Sperm Inside the Queen are Damaged or Dead

During extreme weather, the queens may not die, but the sperm they contain may be damaged. One noteworthy example is the damage done to sperm when bees are exposed to extremes of either heat or cold. While the queen is still alive, all or most of the sperm stored in her body are killed. Once installed into the colony the larvae will all be unfertilized and result in drones. Should this happen, contact your shipper for a replacement queen— they anticipate some loss. Small-scale shippers need to monitor packages for extremes in temperature by using sensors.

Mid-summer queen replacement or supersedure cell in a package bee colony.

The Queen Never Started Laying

After a trans-continental trip, package queens may require three to seven days before they lay eggs. Queens that have been in transit a shorter period of time, properly fed, and installed on drawn comb often start laying eggs sooner.

It is rare, but sometimes queens do not start to lay in the new colony. After one week without eggs, contact your package shipper for suggestions about how this may be resolved. Discuss the possibility of there being a second queen in the hive.

Not Realizing There is a New Queen in the Hive

Quite often, new beekeepers who have not found eggs or brood in their hive end up ordering a new queen. They install a second queen, but the bees killed her! On further inspection, the beekeeper discovers that there already was a new queen, a natural queen produced by the bees, that has started to lay in the hive. This is why I recommend all beekeepers wait for four days before releasing a replacement queen. If present, she may have eggs present.

The Queen Started Laying but Then Queen Cells Appeared

About one third of the packages I have worked with replace their queen during the first season. Since you have had a laying queen and brood, you know that the producer and shipper probably did their job, but the queen still failed. This seems to be one of the commonly accepted outcomes of using package colonies. Your options are to let the bees replace the queen with a daughter or to obtain a new queen at your cost and introduce her into the colony. If you have multiple colonies, you can add a ripe (about to emerge) queen cell from another colony and hope she is successful in mating.

NUTRITION AND OTHER ISSUES

Package bees should be fed in transit and as soon as they are installed. Sometimes package colonies are in a hive all season but never produce enough bees to cover two frames of brood. If the weather has been horrible, you might be correct in blaming the weather. The real problem was poor nutrition, and the beekeeper should feed. If you did feed, and you still have poor buildup, you may have a queen with a blockage in her median oviduct, and she is unable to lay as many eggs as a healthy queen. This is an

New beekeeper in a class inspecting a brood frame.

accident of nature. Replace the queen as soon as you can so the colony builds in strength for the winter.

It is often a painful education to review the mistakes and misconceptions beekeepers have about packages. For example, bee inspector Kim overheard a wife speaking to her husband as he returned home after picking up his first package of bees. "What do you mean your are letting them out? Aren't they like an ant farm?"

We also see how many people are procrastinators or just afraid of the bees. One person asked the package supplier this:

"I did what you said and installed the bees in the hive back in May. Now that it's July, should I light my smoker and look inside?" What will they find? A booming colony with swarm cells? Or a colony in a critical state? Why would anyone start a hive of bees and wait two months to check them?

The opposite extreme is the beekeeper who has opened their new package hive EVERY DAY since installation. Clearly infatuated with their bees, the daily visits were far from quick peaks. They consisted of a full blast of hot smoke, frame-by-frame inspections and extensive photography of what the bees were doing to share with the friends and family. They failed to understand that each

Jars of sugar syrup ready to be placed on colonies that need to be fed.

of their visits were hard on the bees, causing them stress and interfering with colony foraging, brood rearing, egg laying and everything else bees need to do.

Failing to Feed the Bees

Beekeepers look at the bees flying in and out of their new package colonies and figure that the bees are doing a great job and do not need any sugar syrup as food. What they fail to understand is just how much food a new colony needs to obtain even a minimal level of productivity. They might count the number of bees flying out of a hive at different times of the day to see how many foragers are looking for food. Compare that with the number of returning foragers that carry pollen on their hind legs. These are activities the new beekeeper can do that feed the infatuation without disturbing the colony. Sit to the side of the hive and put on your veil while you do this.

When a beekeeper installs a package on foundation or starter strips, the bees carry just a few ounces of sugar syrup inside their honey stomachs. Maybe this is just enough to keep the bees alive for a day or less. The bees know what to do, and immediately start foraging for nectar and pollen and return to the hive. The foragers offer up their nectar to their hungry hive mates, who eagerly take it down and start producing honey and beeswax. It takes a lot of trips to the flowers to obtain nectar, and many pounds of honey to produce beeswax honey comb. Colonies need comb for food storage, but, just as important, without comb the bees do not have a place to raise brood. The clock is running and the bees need a lot of nectar to grow properly and grow the colony.

Gallons of sugar syrup, in a 1 part water to 1 part sugar ratio stimulate the bees to convert the sugar into beeswax by digesting the carbohydrates and secreting wax beeswax scales on glands on the underside of their abdomens.

The bees chew these scales with saliva and shape them into beeswax comb. When the beekeeper provides enough syrup, in a day or two the bees will build several combs with the new beeswax they produce. But this happens only when there is plenty of nectar to gather or sugar syrup to consume.

When the weather is wet or cold, or both, the bees will not be able to forage for nectar. This is when abundant sugar syrup is so essential to the development of the hive. Plus, the syrup will be consumed 24 hours a day until the container is empty. Then the beekeeper needs to feed more syrup.

Because new colonies need a great deal of stored resources to survive through their first year, continue feeding all season when necessary. Especially if the weather does not cooperate and the flowers fail to produce a surplus of nectar. Once the bees have built the equivalent of three deep hive bodies of comb, and filled the combs with honey, then feeding may be stopped until fall, when winter preparation begins. Do not be surprised to find that a hive filled with stored food in mid summer becomes a hive filled with abundant bees and many frames of brood a month later.

Frame containing bee bread, stored pollen.

By November, beekeepers should stop feeding sugar syrup and switch to feeding with frames of stored honey, dry sugar or add honey candy (fondant). Sugar syrup freezes and the bees cannot break cluster tor each it. Always make sure the food is always just above the winter cluster.

Protection

New beekeepers should put on a bee veil (and bee suit and gloves) and light a super **every time** they open a hive. It is embarrassing to see an experienced beekeeper let out a string of profanity when stung because they failed to put on the veil or light the smoker.

For new beekeepers the choice is obvious: Since you do not know what you are doing or necessarily how to do it, you should put on the veil and use a properly lit smoker whenever the hive is opened. No exceptions.

Ignoring Bee Space and Proper Frame Placement

My friend Walter was a careful and studious new beekeeper. He followed the bee club's speaker's comments in tremendous detail.

Walter holding a 'double' frame of brood after following speaker's instructions. A space was open when this frame was moved up, and the bees built on to it.

The speaker said to move a frame of brood into the second box of foundation and placed the box on top of the first. This would stimulate the bees to expand upwards, the speaker claimed. But when Walter came back a few weeks later, he found he had one frame that was two frames deep.

The speaker did not mention that you need to fill the space created when you move a frame with an empty frame of foundation or a starter strip. It may seem obvious, but new beekeepers need to be reminded. The resulting frame, a bit awkward to remove from the hive, was twice as deep as a single deep hive body, and was filled with brood. The bees had really gone to town and built an excellent comb. The resulting frame was amazing but not functional.

Walter and I carefully cut the extra comb off the original frame and tied it into an empty frame. We removed the foundation out of a new comb to do this. We were in a woodlot without any string or wire, so I found some long grass stems to tie the frames into

Bee shaping comb with her mandibles.

A two package hive operation in Kansas. R. Burns

the comb. It worked. The bees attached the comb to the frame and Walter and his bees did not lose any brood.

Bees respond to empty spaces by filling it with comb or propolis (resins obtained from trees). New beekeepers have a lot of trouble determining how much space is too much or too little. But you never take out a frame and fail to replace it with another comb or foundation or a starter strip.

When a space is too small, the bees seal it up with propolis. If it's a wide space, they build additional comb. This precise gap is ideally 3/8 of an inch. If you fail to observe the bee space, you too may be looking for long stems of grass.

Start with Two or More Hives

As kids, my brothers and I had one hive setup and lost the bees every winter. Looking back on our adventures of the 1950s, I am pretty sure we had no idea why the bees died. I blamed the pesticides Dad used on the apple trees, like DDT and parathion. But the colonies could have died from starvation, queen failure or disease. We did not have mites to deal with back then. Now we do.

I recommend that new beekeepers start with at least two colonies of bees when just getting started. Four might be ideal, as it will slow the beekeeper many things. Having multiple colonies provides

1. A frame of brood if one colony is queenless, allowing them to raise a new queen from young larva from a sister hive.

2. A frame of honey if one colony is low on food.

3. Queen cells from swarming or supersedure colonies.

4. Comparison of a good colony with one that is failing, leading to real learning by the beekeeper.

Starting with two or more colonies gives you a chance to see different things happen with different colonies. You could start with one package and one nucleus and compare the two. Two of each would be a better comparison.

Removing Honey Too Early or Taking Too Much

Based on what I have written elsewhere, you may know that I think you should leave most of the honey on the hive during the first year of a package bee colony's existence. You want them to have plenty of food going into winter, and you can harvest extra honey in the spring.

Yet new beekeepers can become greedy. They may take off a box or more of honey and brag about how productive their bees are. They may fail to feed the bees so they survive the winter. They better place their order for a replacement packages early to replace the bees that die over the winter when they run out of food.

It is a tough economics discussion. Package bees are not cheap, and honey is valuable. Sugar is relatively inexpensive, but needs to be mixed with water and fed to the bees. That takes time and effort.

If the bees in the hive are mixing honey and stored sugar syrup together, they create an unintentionally adulterated product that is not pure honey. It should not be sold. Use it for bee feed.

Fermentation

Honey ferments when it contains high moisture, over 19% water. There are many stories about beekeepers who harvested their honey while it was "too set", with moisture content over 19%. When stored in buckets, the lids blew off from fermentation. When stored in jars, they exploded. Hopefully nobody was hurt.

Honey Granulates Naturally

Honey granulates. It is the nature of its sugar and water chemistry. New beekeepers see granulated honey and remove the combs and destroy then. That is extremely wasteful.

Inside the hive, granulated honey is often consumed by the bees by adding water and liquefying the product. Usually, the bees consume the honey before it granulates. I use frames of granulated honey as feed, but place the frames below the brood nest so the bees are stimulated to remove the granulated honey and use it for food.

Varroa Requires Sampling and a Plan of Control

Nobody keeping bees in North America can ignore varroa mites. They may not kill your hive at first, but, eventually, they probably will. Some of us try to keep colonies headed by queens that have proven tolerance against the mites. Hygienic and grooming behaviors are two examples of tolerance behaviors used to minimize mites in bee colonies.

These traits may slow the reproduction of the mites. But in the fall, in many areas, something called the varroa bomb goes off. The bomb refers to the mites from your neighbors hives that are moved by bees leaving the neighbor's dying hives or by your bees robbing out the weakened hive. Both result in increased mite numbers in your colonies. The varroa bomb is evident when your mite counts go from one mite per 100 bees to 25 mites per 100 bees in the matter of a few weeks. This is usually in the fall.

Sample your colonies several times a season. I recommend once a month. In many areas, the most important sampling periods are from August to November. The powdered sugar sampling system works well and does not kill the bees.

Once the number of varroa mites in a hive reaches two per 100 bees, you need to treat somehow. Know what you will use and how to use the compound. Use one of the essential oils or organic acids for your initial control method. Spring and fall treatments are most common. Conduct another sample after you follow the treatment procedure and see how well you have done.

For detailed information on varroa mites, check the *Honey Bee Health Coalition's Tools for Varroa Management (A Guide to Effective Varroa Sampling & Control)* at honeybeehealthcoalition. org.

Powdered sugar and a shaking jar equipped with screen mesh are used to sample mite populations without killing bees. It is often considered 90% as accurate as the solvent-based sampling method shown on the next page.

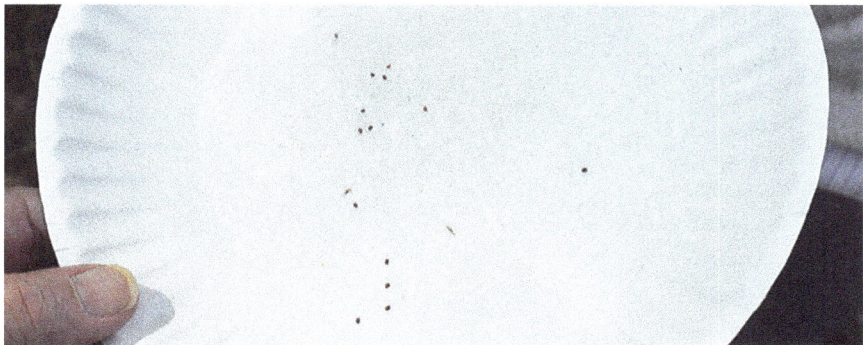

After bees are shaken, mites dislodge and fall when the jar is shaken over the paper plate. Water mist liquefies the sugar, revealing the mites for a count.

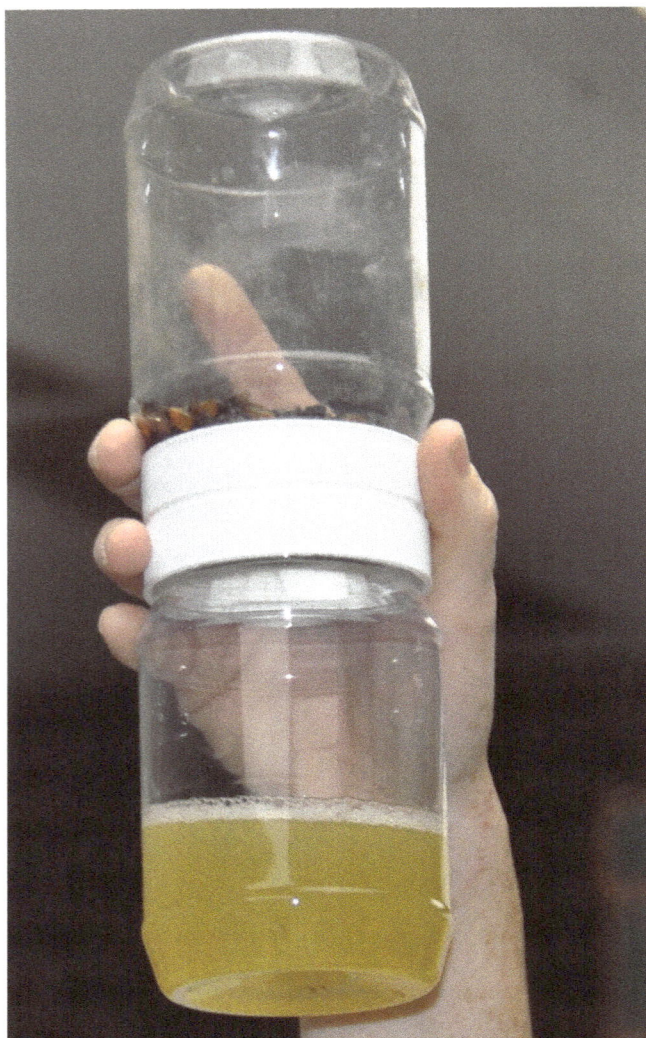

This double shaker jar places bees in the upper jar, allowing the release of mites from the bees' bodies. Bees die in the sampling method (L). Mites removed from the bees' bodies (R).

—INDEX—